Genetic and evolutionary diversity

Genetic and evolutionary diversity

The sport of nature

Second Edition

L. M. Cook
The Manchester Museum
University of Manchester

and

R. S. Callow
School of Biological Sciences
University of Manchester

Stanley Thornes (Publishers) Ltd

First edition published by Chapman & Hall in 1991

Second edition published in 1999 by:
Stanley Thornes (Publishers) Ltd
Ellenborough House
Wellington Street
Cheltenham
Glos.
GL50 1YW
United Kingdom

99 00 01 02 03 / 10 9 8 7 6 5 4 3 2 1

A catalogue record for this book is available from British Library

ISBN 0 7487 4336 7

Typeset by WestKey Ltd, Falmouth, Cornwall
Printed and bound in Great Britain by T.J. International Ltd, Padstow, Cornwall

Contents

PART TWO POPULATION PROCESSES

Preface to second edition

The elementary evolutionary process is change in gene frequency. From so simple a beginning, endless forms most beautiful and most wonderful have been, and are being evolved. These statements, of Sewall Wright and Charles Darwin, respectively, span the range of ways in which evolution may be viewed. In this book, we examine three approaches, population genetics, cytogenetics and population ecology in order to present some themes relating to generation of organic diversity.

Population genetics has its origins in the late nineteenth century desire to understand the nature of inheritance and the process of evolution. The subject developed as an interplay between field observation and laboratory experiment, on the one hand, and of mathematical theory, on the other hand. Observation stimulated theoretical analysis but theory often determined what the experimenter looked for, and has sometimes outstripped the data to which it should relate. Population genetics probably works best describing the minute behaviour of alleles at single loci. It is often used to solve practical problems. Examples are the measurement of inbreeding in relation to inherited diseases, development of strategies for insect pest or disease vector control, assessment of the consequences of release of genetically modified organisms. In the course of such investigations, surprising amounts of genetic variability have been detected at all levels from phenotype to DNA sequence.

Cytogenetics depends on technical development and interpretative skill. It is concerned with the way in which genetic information is transferred from one generation to the next. It is apparent, however, that the processes involved include not only the mechanics of transfer but also optimization of the balance between genetic conservation and novelty.

Evolutionary biology is field oriented, verbally expressed and concerned with the complex interplay of environment, inheritance, ontogeny and most importantly, of history. For a long time, genetics and evolutionary studies developed independently, with wary glances at each other and at the ecological principles on which both are based. Although we now tend to look at population biology as a whole, there are still doubts as to how well the micro-evolutionary level can be related to major long-term evolutionary changes.

This book began as a short basic introduction to population genetics. It was intended to accompany a course for undergraduates with a background in genetics and zoology but little mathematics. Essential features of the theory are outlined. The few examples employed are mostly classical ones used as models to relate to the theory. No attempt is made to review the extensive field evidence or the more advanced theoretical developments. At the genic level, the treatment relates to animals. The important area of quantitative inheritance is now given more attention, and a variety of investigations of plants allow breeding systems and the significance of chromosome reorganization in evolution to be considered.

The first and last sections arise from a question we have asked in connection with teaching: does an analytical approach centred on single-locus, and occasionally multi-locus, models really permit insights into the larger evolutionary process? One way to approach the question is to consider the inordinate variety of species, a diversity which parallels the genetic diversity found within populations. The first three chapters are a glance at aspects of ecology and systematics and ask some questions about species diversity which might be expressed in terms of population genetics. The final section is an attempt at a summing up. It is, of course, both a personal view and a rash undertaking. One conclusion is that comprehension of many aspects of evolution and species diversity requires further study of adaptive constraint. Population genetics can contribute through the study of locus interaction. Another is that evolution at the genic level cannot be properly understood without considering the way transfer of inherited information is accomplished. Selection on genes influences structure of the genome. Patterns of chromosome transfer affect genic balance and phylogenetic evolution.

An eventful decade has elapsed since the first edition appeared. Many important investigations of particular species or systems have been carried out, the meta-population approach has become standard, comparative methods of phylogenetic analysis have been developed and are routinely applied to molecular data. There has been a move from analytic to simulation modelling, to deal with more complex theoretical problems. There is hope for a human diversity project to complement the human genome project. Molecular population genetics is now firmly established.

Is there, then, still a place for a book such as this? We think there is. Evolutionary genetics must consider all levels, from the gene through the genome to the phenotype. Current research rests on past foundations, and requires a framework of whole organism biology. The neutral theory has lost some ground over the last two decades, but its validity in fields such as molecular phylogeny should be debated in relation to alternative models and evidence from empirical studies. Students have gained much knowledge of newer fields, sometimes at the expense of perspective. In this book we present ideas and develop a theme, rather than review recent field and laboratory research. Each chapter ends with a summary, representing a view of the subject. Are these statements true? Are they sufficient? Are we any the wiser? These are questions readers can test against their own interpretations of the data.

Finally, a note should be added about our use of references. The text naturally reflects what we have absorbed from the literature, but for the most part citations of original sources have been left out in the hope of improving readability. References cited (usually in figures and tables) are given at the end, together with a bibliography which provides an introduction to the areas of population genetics, dynamics and evolution discussed. This list goes back no further than 1970 and deals with both theory and real organisms; it should provide points of entry into the extensive literature. Appropriate titles in the bibliography are indicated at the end of each chapter.

PART ONE
Species abundance

Species diversity 1

1.1 PROLOGUE

> Up to this time the industry of men has been great, and very curious in marking the variety of things, and explaining the accurate differences of animals, herbs and fossils, – the chief part of which are the mere Sport of Nature, rather than serious and of use towards the sciences. Such things tend to our enjoyment, and sometimes to even practical use; but little or nothing towards an insight into Nature. And so our labour is to be turned to inquiry into, and notice of, similarities and analogies, both in the whole and in the parts of things: for these are they which unite Nature, and begin to establish sciences.

This passage is from Francis Bacon's *Novum Organum or Directions Concerning the Interpretation of Nature*, Book 2, 27. It was quoted in the present translation by a little-known Victorian naturalist, Thomas Vernon Wollaston, in a book he wrote in 1856 entitled *On the Variation of Species*. Wollaston was born in 1822 and died in 1878. He lived for some years on Madeira, and made significant contributions to our knowledge of the fauna of that and other Atlantic island groups, especially in relation to their insects and molluscs. He published a book on Madeiran insects in 1854 and a monograph on the terrestrial snails came out shortly after his death. His life spanned a period when scientists turned from looking for the ways in which individual plants and animals embodied common features defining the species to which they belonged to studying the intrinsic variability of species. Darwin's insight was that variation within species leads to natural selection and permits evolution, whereas earlier scientists followed the principle of seeking the common features which identify the group and ignoring the divergences. The quotation from Bacon was an affirmation of the Platonic idea of species, which went with a belief in their individual creation. Wollaston dedicated his second book to Darwin, 'whose researches, in various parts of the world, have added so much to our knowledge of zoological geography'. He disagreed with him strongly, however, over the question of evolution, and in later life attacked him anonymously in reviews. Darwin wrote to a correspondent that he could recognize his attacker from his prose style, which, even for a Victorian, was contorted.

Although Wollaston had religious views on the delineation of nature, the stated basis of his objection to evolution was not a religious but a scientific one. He had observed the diversity of species, both in one place and between different geographical races. *On the Variation of Species* was an enquiry into the factors which correlated with, and in some sense could be said to cause, this variation. Nevertheless, he maintained, there was no evidence that these factors ever led to intergradation from one species to another:

> It does indeed appear strange that naturalists, who have combined great synthetic qualities with a profound knowledge of minutiae and detail, should ever have upheld so monstrous a doctrine as that of transmission of one species into another, – a doctrine, however, which arises almost spontaneously, – if we are to assume that there exists in every race the tendency to an unlimited progessive improvement. There are certainly no observations on record which would, in the smallest degree, countenance such an hypothesis.

This view is not the result of a failure to study nature; his application was exhaustive and minute. Wollaston mentions two types of evidence which he considers expose the absurdity of the transmutation hypothesis. On the one hand, we do not find intermediates between species. In his last book, the *Testacea Atlantica* (1878), he devotes more than 500 pages to detailed taxonomic descriptions of land molluscs which, he claims, show that species are all distinct from each other. It is hard to find the central message of this book full of particularities, but at one point (p. 561) he states:

> these remarks are by no means intended to insinuate that the lines of demarcation between species, when correctly interpreted, are ever, in my opinion, really confused or doubtful (the exact opposite having always been my firm belief); but for us to determine them aright is quite another matter, and I am willing therefore to admit that we may often be seriously mistaken in our endeavours to decypher them.

With very few exceptions his species are accepted today.

The second piece of evidence is that we frequently find groups of similar congeneric species living side by side in the same habitat. Wollaston gives as an example 18 species of Madeiran beetles which are quite distinct but all closely adapted to the particular habitat in which they live. If evolution adjusts species to their habitats then there should be only one species in this particular habitat, or a group of species each fitted to different conditions. Thus, evolution does not explain the patterns, and we have to account for the amazing diversity of nature in other ways. We are now accustomed to the idea that the theory of evolution does explain these phenomena, but it is salutory to consider the evidence studied so critically by such men as Wollaston. The idea of evolution by natural selection requires that one species transmutes into another. If we do not find evidence for transmutation, why not? The theory does not require, however, that groups of similarly adapted and closely related species should coexist. Do they, and if so, how does this pattern come about? Nineteenth century naturalists were not fools and they had plenty of observations at their disposal. Darwin's contribution was outstanding just because nature is enigmatic. The theory of evolution by natural selection required a profound insight, and we cannot understand all the patterns in nature as necessary consequences of a dynamic evolutionary process.

For those who accept a synthetic theory of evolution this is a challenging position, and the study of nature raises questions for which answers can be attempted in the light of modern knowledge of genetics and ecology. However, there are still people who deny the evidence for evolution and write books about their views, as there have been since evolution was first discussed. There is also a more subtle tendency, based on a misunderstanding of the synthetic approach, which seeks to explain evolution by describing the conformity between structure and environment, without the agency of natural selection. Variety and diversity are central to our perception of living things. It is fascinating to examine the variation itself, how it arises and what it tells us about evolution.

1.2 MOLECULAR CURRENCY

The Earth has been in existence for some 4.7 thousand million years. The earliest evidence of life can be dated to about 3.2 thousand million years ago; there was probably a considerable range of bacterial types in existence a thousand million years later. In the Cambrian period which goes back to about 600 million years, marine algae, molluscs, arthropods and brachiopods flourished. Some of these are recognizably like modern forms. At one level, living systems are highly conservative. In the whole of the recorded sequence life has been replicated through a system in which trios of the same four bases code for the same 20 amino acids. In functioning genes the same terminator codes are used to stop transcription, and very similar initiator sequences are used to start it. All living organisms are based on a cellular structure, with the cells falling into one of two general patterns, the prokaryote, found in Archaeobacteria and the various types of bacteria, and the eukaryote, found in all the rest. Functioning of the cells depends on a common energy-carrying molecule, ATP, through glycolysis, so that the enzymes involved are present in all living things except some bacteria. All aerobic cells operate the citric acid or Krebs cycle, and therefore have another set of enzymes in common. Solar energy trapped and converted to chemical energy by living organisms is obtained only through photosynthesis, without which the whole marvellous system of life would run down.

When we look at the detailed sequences of DNA carried by different organisms, common structures and functions are often found to be coded by genes which are extremely similar in very different forms of life. Sometimes homologous genes and their products are used for different purposes in different kinds of organism. Myoglobin and haemoglobin are recognizably like each other in structure, to an extent which makes it possible to estimate the time at which the two genes had a common ancestor, a single type of molecule present in some creature 800 million years ago. Thyroxine is a metabolic hormone in mammals and it controls metamorphosis in frogs. Prolactin generates milk production in mammals and crop secretions in birds. It is involved with determination of water-finding behaviour in amphibia and with salt balance in fish. Alkaloids which have a protective function in plants are used in courtship by butterflies. Proteins involved with control of encystment and sporulation in microorganisms are related to crystallins of the vertebrate lens. The vertebrate proteins may be derived from ancestral forms with a role in stress response, possibly osmotic stress. In all these cases the structure of the molecules is conserved while the function varies to suit different patterns of development or organization.

These types of evidence show that biological diversity is achieved by ringing the changes on quite a limited chemical repertoire. We do not understand the constraints which dictate that life should be founded on the four-base nucleic acids and 20 amino acid polypeptides. There may be some fundamental reasons why alternative polymers could not replicate and form the basis for development of different energy-generating systems. These forces seem to have operated to ensure that the molecular structures have remained essentially constant over hundreds of millions of years. When we look at the differences between individuals, however, it is apparent that while the molecular story is notable for its conservatism it can nevertheless result in an immense variety of phenotypes within a species and of species in a habitat. It is as if a simple currency had been devised to cater for a multiplicity of transactions.

1.3 SPECIES MULTIPLICITY

Living organisms may be divided into five kingdoms, depending on features of their cell structure and metabolism. These are: the Monera, prokaryotic and mostly bacteria; the Protista, single-celled and, like the rest, eukaryotic; the Fungi; the plants; and the animals. This system succeeds a three-kingdom classification, and has itself been criticized and revised. Further changes will be made as knowledge develops (e.g. see Williams and Emberley, 1996). Whatever the eventual consensus, each kingdom is divided into major phyla. Table 1.1 lists these with an estimate of the number of described species in each.

There are more phyla than most biologists can remember and in them a truly enormous number of species. The animal kingdom is the most species rich, with 1 110 000. A long way behind come the plants, with 290 000, the fungi with 60 000–90 000, the protists with about 40 000 and the Monera with 4000–5000. Most plants are dicotyledons (200 000). Monocotyledons amount to 50 000 and the conifers or gymnosperms come well behind at a mere 700. The majority of species are arthropods (900 000), and of those the majority are insects (750 000). The majority of these, in turn, are beetles (300 000), which therefore rank in terms of species number as the most successful group of organisms. About half the chordates are bony fish; there are fewer than 5000 types of mammals. After the arthropods, the largest phylum is the Mollusca, with 100 000 species, a fact which for most people is unexpected. The distribution of species number between phyla is strongly skewed, with some having only a handful of species and others having tens of thousands. It becomes more symmetrical if we take the logarithms of the numbers, a pattern which is also seen when the numbers of individuals within species is considered.

The numbers given above are numbers of described species. It is certain that not all species have been recorded, but can we estimate the likely number which remain undescribed? A few years ago the answer would probably have been that biologists have got about halfway. In the late 1970s the entomologist T.L. Erwin carried out surveys of the insects, especially beetles, of South American rain forest using the very efficient method of fogging the forest canopy with ultrafine droplets of insecticide. The method was to reward him with many times the number of species that had hitherto been observed. Since rain forest is known to be extremely species rich and is very poorly studied, it is not unreasonable to extrapolate from his finding to other groups of organisms and suggest that we may only have begun the process of stock taking. Estimates of between 10 and 80 million species have been put forward on the basis of this revised view, a reasonably conservative guess being about 20 million. Why on earth are there so many species?

Before attempting to answer this question, it is necessary to consider

Table 1.1 Numbers of described species in the kingdoms and phyla of organisms

Kingdom Monera	
Schizomycophyta	1500
Cyanobacteria	2500
Kingdom Protista	
Superphylum Protophyta	
Chrysophyta	6000
Cryptophyta	100
Pyrrophyta	1000
Chlorophyta	6500
Euglenophyta	500
Superphylum Protozoa	
Sporozoa	3600
Sarcodina	17 500
Zoomastigina	1100
Ciliophora	6000
Mesozoa	50
Kingdom Fungi	
Myxomycetes	500
Phycomycetes	1300
Ascomycetes	16 000
Basidiomycetes	16 000
Deuteromycetes	10 000
Kingdom Plantae	
Rhodophyta	3000
Phaeophyta	1500
Bryophyta	23 000
Psilophyta	4
Lycopodiophyta	1200
Equisetophyta	25
Pteridophyta	10 000
Gymnospermae	700
Angiospermae	250 000
(Dicotyledons	200 000
Monocotyledons	50 000)
Kingdom Animalia	
Poriphera	10 000
Coelenterata	9000
Ctenophora	90
Platyhelminthes	12 000
Nemertina	800
Rotifera	1500
Gastrotricha	170
Kinorhyncha	100
Acanthocephala	500
Nematoda	10 000
Nematomorpha	200
Entoprocta	70
Bryozoa	4000
Phoronida	20
Brachiopoda	250

Mollusca	100 000
Priapulida	8
Sipuncula	250
Echiura	150
Annelida	8000
Arthropoda	900 000
(Insects	750 000)
Chaetognatha	50
Pogonophora	80
Echinodermata	6000
Hemichordata	90
Chordata	45 000
(Bony fish	20 000
Amphibians	2500
Reptiles	6000
Birds	8600
Mammals	4500)

Based on several sources, Altman and Dittner (1972) and Whittaker and Margulis (1978).

how meaningful the numbers are and what criteria are used to establish them. First, the numbers in different groups must to some extent be a function of our interest in that group and the ease with which it can be studied. Insects are numerically abundant, often closely associated with human beings and injurious to them, and they are easy to characterize on the basis of external features. It would not be surprising if they were relatively overrepresented in the list. Birds receive a great deal of attention, coelenterates comparatively little; the number of recorded species of birds may be an overestimate, that of coelenterates is certainly an underestimate. It has only recently become possible to study the cyanobacteria in the oceanic plankton because of their small size and inaccessibility. It is probable that their species number is greatly underestimated. The figures in Table 1.1 probably give a reasonable guide to the relative number of species in different groups although they are bound to be subject to inaccuracy arising from differences in attention accorded them and in ease of study. Another type of question to ask is whether they really represent distinct categories or are merely names applied for convenience to sections in a continuity of different forms. In order to attempt to answer these questions we have to consider briefly the methodology used to establish them.

1.4 METHODS OF CLASSIFICATION

The decision to create the categories discussed above involves judgements about the relative importance of different attributes of organisms. The kingdoms used here have not always been employed. At one time the

Cyanobacteria, which are photosynthetic were considered to be plants. The Protozoa, which do not photosynthesize, were included among the animals and separated from the photosynthetic single-celled protists, which got into the plant kingdom alongside the algae. The present classification draws attention to some important features of organization, but it would not affect the argument if the phyla were to be grouped in a different way. If the number of species in each group is no more than a convenient artefact, however, then no problem is associated with their abundance. Why do we accept the species as a fundamental category? In order to try to answer this question we have to examine the procedures used to classify organisms. These involved two processes: taxonomy, the application of rules for naming the categories, and systematics, the process of classification itself. We very often use these terms loosely, and call the classification process taxonomy, but strictly speaking the two operations should be distinguished.

In the first attempts to sort organisms into categories they were arranged in order as they differed from man. Aristotle divided animals into vertebrates and invertebrates. In the vertebrates there were seven classes: (1) man, (2) cetaceans, (3) viviparous quadrupeds, (4) birds, (5) amphibians and reptiles, (6) serpents and (7) fish. The invertebrates consisted of five groups: (8) cephalopods, (9) crustaceans, (10) arthropods, (11) molluscs other than cephalopods, echinoderms, etc. and (12) sponges and coelenterates. As one proceeds along this sequence the mode of reproduction changes from internal fertilization and viviparity in man through egg laying and external fertilization as far as category (12), which Aristotle assumed to be the products of spontaneous generation. Although we would make several changes to his categories, such an arrangement is not entirely unfamiliar to us today. It immediately introduces another theme. We are unique among living things, at least in our ability to reason, and the category most distant from us consists of animals with very simple organization, so perhaps the classification is a hierarchy ordering animals on a scale from advanced at one end to primitive at the other. Such an idea of progress used to be accepted as fact. It is recorded in Genesis and recognized in the allocation of a soul to man alone. Biologists now tend to avoid this conclusion, and argue that all kinds of organism are highly adapted to their own walks of life – lack of complexity does not in any sense imply imperfection. But is that really true? It is worth keeping the question of degree of perfection under review when considering the classification of organisms.

To take examination of the process of classification further it is interesting to consider the work of two outstanding biologists of the eighteenth century. Georges Cuvier was born in France in 1769 and died in 1832. He had a profound influence on the development of zoology, being the first to establish systematic comparative anatomy and to recognize the interdependence of parts of the body: 'The organism forms a connected unity, in which the single parts cannot change without bringing about modifications in the other parts'. He collected and dissected, and studied both living and extinct forms, for the first time linking palaeontology with comparative anatomy. Although his researches led him to oppose the theory of descent he recognized the existence of four major types of organization. These were the vertebrate, divided into Bimana, Quadrimana, etc., the molluscan, which included barnacles and brachiopods, the articulate, including annelids, insects, crustacea and arachnids, and the radiate, including echinoderms and coelenterates. The differences were supported by embryological research and the classification was published in 1817 in *Le Regne Animal, distribué d'après son organisation*. Like Aristotle's classification, this one proceeds from complex to simple with respect to nervous system, sense organs, methods of respiration, circulation and so on. From his experience of comparative anatomy, Cuvier also claimed for his system that it was natural, as distinct from the artificial classification of flowering plants employed by Linnaeus.

Carl Linnaeus, or Linné, had an interest in botany from his early youth. He studied medicine at university but became assistant to the Professor of Botany in Uppsala in 1730, at the age of 23. He lived until 1778. His work took him on exploring and collecting trips, to Swedish Lapland among other places, and to visit and work with eminent botanists in several countries of Europe. He published several books, of which the most famous now is the *Systema Naturae* (1735). He insisted on the fixity of species and contested the belief in spontaneous generation. The *Species Plantarum* came out in 1753, and in it he first fully established the custom employed today of using a second, specific, name as well as the generic name to identify a plant or animal. His system for classifying plants was deliberately an artificial one, based on their sexual characteristics – the number and arrangement of stamen, ovary, stigma and style. This approach worked well as a means of categorizing plants so that they could be referred to and arranged in collections, but it did not address itself to the question of interpreting them as varying manifestations of a pattern of life. It is not a natural classification, in the sense of Cuvier, because it is based on one set of characters rather than characters from all parts of the organism. What insight do we gain, then, from use of a natural classification?

The advantages, and difficulties, can be seen by deliberately creating a very diagrammatic problem. Suppose we wish to construct a list showing the characteristics of birds and mammals. It would contain features such as the following.

1. Bird: internal bony skeleton – Mammal: internal bony skeleton.
2. Bird: oviparous – Mammal: viviparous except for monotremes.
3. Bird: feathers but not fur – Mammals: fur (usually) but not feathers.

4. Birds: do not suckle young – Mammals: suckle young.
5. Birds: no teeth – Mammals: teeth usually present.
6. Birds: heart and major arteries avian type – Mammals: heart and major arteries mammalian type.
7. Birds: usually able to fly – Mammals: usually unable to fly.
8. Birds: limited size range – Mammals: greater size range.

Of these characters, (1) is useless for separating the two groups because it is possessed in common by them, and that would go for any other shared features we chose to add. The first principle in systematic procedure is therefore to look for distinctive, rather than shared characters. Of these, several, such as (2), (3), (5) and (7), are fairly reliable but not entirely so, because in one or other group there are exceptions. Some mammals fly, and not all birds do so. Although many mammals are much larger than most birds, body size in the two groups overlaps. A logical next step is to look for characters which always distinguish the two groups, known as diagnostic characters. In this case, they are the suckling, or not, of the young and the arrangement of the arteries running from the heart. Birds and mammals both have complete separation of oxygenated and deoxygenated blood in the heart, so that pulmonary veins bring the blood from the lungs into the heart, from where it is pumped into the dorsal aorta. In birds the artery joining the heart to the dorsal aorta passes on the right side of the body, while in mammals it passes on the left. In amphibia there are arteries on both sides, and the avian and mammalian arrangements represent loss of branches and restriction to one side of the body or the other. How do we know these characters are diagnostic? Because they are always different in the group called birds and the group called mammals. The two groups are distinguished, however, by some sort of averaging process which makes allowance for the occasional exceptional features. The averaging process comes naturally to us when the groups are as well known and distinct as birds and mammals, but is much more difficult when they are unfamiliar. Diagnostic characters usually emerge in the course of the averaging, rather than being apparent from the start. If we simply wish to catalogue museum specimens, a distinction such as body size might be very sensible, because it would indicate the type of drawer or box to use to keep the specimens. If we were interested in practical considerations then categories such as game, vermin, domesticated, etc. could be, and frequently are, used. Categories based on the averaging process are manifestly more natural, however, in that they describe something real about the quality of 'bird-ness', as distinct from 'mammal-ness'.

When trying to undertake an averaging process there are two problems to be considered. One is correlation. In birds, oviparity, the comparatively small size and limited size range and the absence of teeth may all be consequences, to some extent, of the need to fly, whereas the handedness of the artery is not. When they are all included, each of the characters associated with flight is less important in describing a bird than a character independent of flight. It follows that they should count for less, and the idea of weighting of characters is introduced. In an ideal classification, characters should be uncorrelated. When they are correlated, some sort of allowance should be made.

The second problem arises when there is convergence. Birds and bats fly, whales and fish have similar shapes, squids and mammals have similar eyes. The taxonomist has to recognize these similarities as consequences of convergence between otherwise very different forms of life and weight them accordingly. No difficulty is likely to arise with the examples given, but the process of disentangling convergent characters from ones which truly reflect the organization of the group can be difficult and subjective for less familiar organisms. Even if we take newts and frogs, their shared amphibian characters are probably mostly convergent and not indicators of close affinity.

As an example of a real classification we might consider the carnivores. The order Carnivora is a group within the class Mammalia. It is divided into two suborders, the Pinnipedia, or seals and the Fissipedia, which comprises the rest. Among these, there are dogs, cats, weasels, civets and bears, but also animals which diverge further from the central pattern, such as the panda, the kinkajou and the aard wolf, *Proteles*. The Carnivora are classified as having:

1. a relatively reduced dentition consisting of small incisors, large canines and reduced, cutting cheek teeth;
2. limbs capable of complex movement;
3. absence of the clavicle bone in the shoulder;
4. fusion of three bones of the wrist, the scaphoid, lunar and centrale.

Of these characters, the dentition is not always of the kind described; the panda has cheek teeth adapted for chewing and *Proteles* has almost lost its teeth. In any case, judgement of such a set of characters requires broad experience of the Mammalia as a whole before the condition described can be recognized. Some members of the group have characters convergent on those of other groups, such as the flippers of seals or the prehensile tail of the kinkajou, which it shares with opossums and New World monkeys. The absence of a clavicle is part of the set of characteristics allowing free movement to the limbs, so that characters (2) and (3) are really telling us the same thing. Condition (4) is diagnostic of the group, but is the type of character which would only be established after the conception of the order Carnivora had been firmly established.

Most systems of taxonomy aim to produce a natural classification, and they are all to some extent subjective. Ideally, they are based on good diagnostic characters, which members of a group have in common and by which they differ from other groups. In the past these have often been taken to be characters unimportant to the organisms (non-adaptive), and therefore invariant. Good diagnostic characters do not vary with conditions of life, as the limbs of mammals may be adapted for running, climbing, swimming, etc. Nevertheless, they are not so much unimportant and non-adaptive as severely constrained. It is probable that the arrangement of the heart and blood vessels found in birds works with the same degree of efficiency as that of the mammals, so that both groups are equipped with effective and functionally equivalent vascular systems. Any variant arrangement which made one converge on the other, however, would be likely to be much less satisfactory. The amount of variation is constrained by functional needs and the character is therefore invariant and diagnostic.

The arrangements resulting from taxonomy are discontinuous groups until one gets down to a degree of similarity below the species level. Species come out of the process as real entities, having characters in common and distinct from their most similar relatives. These are then grouped into genera, families, orders, classes and phyla up to the level of kingdoms. Taxonomists working in different kingdoms have to use different criteria, but they understand and accept the judgements of their colleagues. Over the past 30 years two trends have arisen designed to put systematics on a firmer basis. In numerical taxonomy, these have set out to automate the process using computers, and in cladistics, to codify the rules.

If we use one character to distinguish between groups it may not be an appropriate one, if a few then they may be correlated or convergent. One way of minimizing this problem would be to score a very large number of characters. Good diagnostic features are likely to be included and effects of correlation will be minimized. In numerical taxonomy, or phenetics, general resemblences are averaged on this principle. The first step is to create a table of characters possessed by each taxon (called an OTU, for operational taxonomic unit). These may be of any kind: presence or absence, metrics, ranks, colour combinations, etc. All tell us something about the similarities or differences involved, although mixing of data types may complicate the analysis. A table something like that in Figure 1.1(a) will be produced, and there should be 5–100 characters represented. All possible pairwise comparisons are then made between the taxa; for n taxa there are $n(n-1)/2$ of these. To make the comparisons economically, it is necessary to have a similarity index, expressing as a single figure the degree of resemblance of any pair. There are many possible ways of calculating similarity, which reflect different features of the resemblance. One approach is to add all the characters which a pair of taxa have in common and divide by the total number of characters they possess altogether. This index will vary from 100%, if each possesses the same number of characters and all are common, to zero, if they differ in every respect. Features which any pair lacks, but which are possessed by other taxa being considered, are ignored. A different approach would be to calculate the correlation coefficient, r, for each pair. This has values varying from + 1, for identity, to − 1 if no character is shared. Shared absences are included in the calculation. The value $r = 0$ also has a meaning, namely that the taxa are as alike as would be expected by chance if the characters of the whole set had been sampled at random. In taxonomy it is usually most sensible to exclude characters possessed by neither of the two groups being compared, even if they are present in some other similar organisms. This

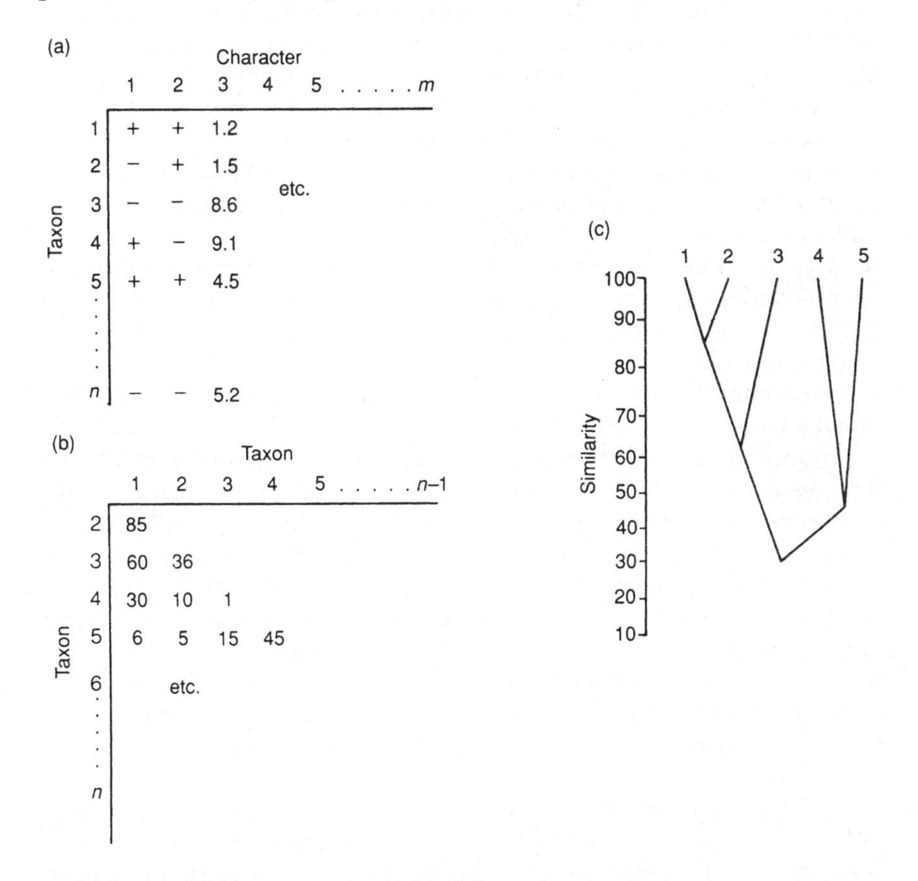

Fig. 1.1 Grouping by overall similarity. The data consist of *m* characters measured in *n* taxa or groups of organisms. These are arranged in table form in (a). A measure of similarity between taxa, such as the correlation coefficient or fraction of characters in common, is used to produce the similarity matrix in (b). This may be summarized in a branching diagram such as (c). The nodes in the diagram may represent the greatest similarity between a group and the next taxon or group, or the average similarity between them: greatest similarity is shown, given numerically in (b). This method may be used for clustering other types of data, such as the species present in different habitats or on different islands, or the gene frequencies at a set of loci in different populations. The methodology is discussed by Sokal and Sneath (1963).

process of measuring similiarities is also widely used in ecology, for purposes such as comparing species composition between habitats, and there the absence of a particular species from two habitats may sensibly be regarded as a feature distinguishing them from other habitats. In a different field a much-used method for comparing genetic variation between populations, Nei's index, is like r except that the mean values for the two samples are set at zero.

Having calculated the values of the similarity index, the investigator then plots them as a matrix of similarities for all the non-identical comparisons, as in Figure 1.1(b). This could then constitute the end of the analysis, in that the matrix may show patches of high similarities, indicating close affinity, separated by patches of low values, indicating that the taxa involved were less similar. A further step which is usually then taken is to plot the data as a diagram of the similarities between groups (Figure 1.1(c)). Here, the taxa have been clustered according to the maximum similarity between a member of a group and the next taxon linked to it (the minimum distance method). An alternative method, which will usually produce a similar picture, is to find the average distance between the members of the groups and the next linked taxon. Once this is done, we have a tree indicating the average degree of similarity of the set of taxa being compared.

The strength of the numerical taxonomic method is its objectivity. Provided a large number of characters are considered the clustering is likely to produce the same pattern whatever set of characters is used, so that the dangers of weighting effects are minimized. On the other hand, redundant information is included, in the form of characters common to several or all of the taxa, which will alter the overall similarity but have no discriminating power. Taxa which come out as similar to each other may have no common diagnostic characters, and must then be recognizable only by their general resemblance. One practical drawback is that few systematists are likely to have the time or patience to collect the enormous amount of data needed for the advantages of the method to be manifest, so that the result will still depend on the judgement used in choosing the characters.

Cladistics is a method of taxonomy intended to assist and clarify that judgement. The name is derived from the Greek for a branch or shoot, and was first used by a German specialist in Diptera, Willi Hennig. He published a book on the principles of taxonomy in German in 1950. This had little impact on the English-speaking world until an English translation of a revised version appeared in 1966. Entitled *Phylogenetic Systematics*, it was taken up by a band of enthusiastic devotees. Every taxonomist, Hennig argued, uses a process of logic when erecting a classification, which should be intelligible and acceptable to other taxonomists. It is therefore possible to establish an agreed system which uses the best procedures, and when this is done, the best classification will be produced. The system is based on three axioms: (1) that shared features manifest a hierarchical pattern, (2) that the pattern is economically expressed by branching diagrams (like the one in Figure 1.1(c)) called cladograms, and (3) that nodes of the cladogram indicate features grouped by the node, so that the cladogram is the classification. An example of the method is given in Figure 1.2(a), which shows a character matrix for four taxa. The letters represent different states of the eight characters scored, while dashes indicate that the character is absent.

There are 15 ways of grouping these taxa in pairs, so that rules are required to decide which is the most appropriate. If the presences and absences are counted the clusters come out as in Figure 1.2(b). Characters 3,4 and 5 group A and B together, while characters 2, 3, 4 and 5 group C and D. Characters 1 and 8 do not fit with this grouping, however, and are

(a)

Taxon	Character 1	2	3	4	5	6	7	8
A	–	G	L	S	D	G	V	L
B	M	E	L	S	D	Q	V	L
C	–	–	–	–	–	A	V	L
D	–	–	–	–	–	T	V	N

(b)

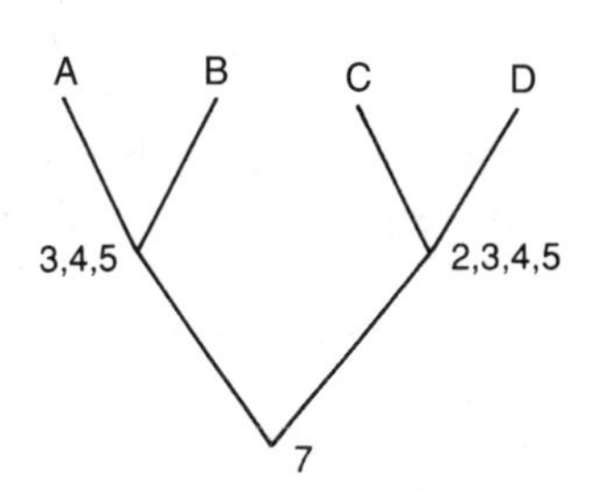

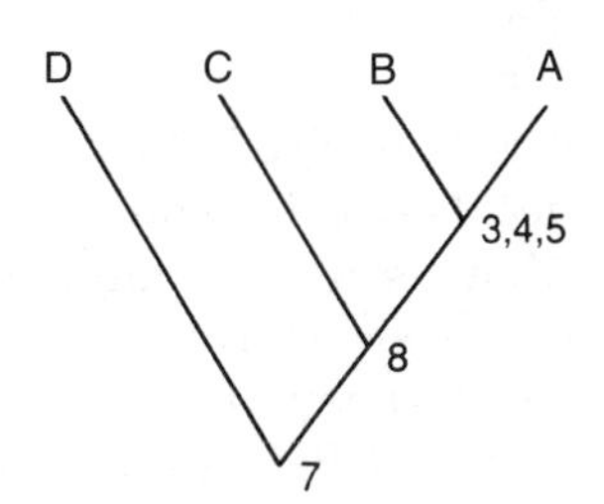

Fig. 1.2 Classification using the cladistic method. The character matrix to be used is shown in (a); letters indicate different states of the character and dashes indicate absence of a character. (b) and (c) show two ways of linking the four taxa. In (b) presences and absences are counted, while in (c) absences are ignored. The rules suggest that (c) is the better classification. Numbers at each node represent the characters shared by taxa above that node. There are fewer incongruent characters, which assort differently between taxa, in (c) than in (b). The characters are amino acids at positions in the myoglobin chain of man, alligator, tuna and a shark, respectively, so that (c) also represents the classification we should adopt on other grounds. Reproduced with the permission of the Institute of Biology from *Biologist*, Patterson (1980).

said to be incongruent. Alternatively, we may consider only characters which are present, and ignore the absent ones. When this is done, the cladogram comes out as in Figure 1.2(c). Here, characters 3, 4 and 5 group A and B together, while C and D are distinct from them and from each other. No characters are now incongruent, and for this reason the classification is a better one. The procedure of cladistics is always to group taxa according to features shared by the group, which distinguish the group from others. A second aim of the method is that it should be predictive, in that a classification based on one set of characters should conform with that based on another. In this case, the characters are amino acids in particular positions in the myoglobin chain, and the taxa are, respectively, man, alligator, a teleost fish and a shark. The grouping preferred therefore supports our judgement of affinity based on other characters, while the first of the two does not.

The intentions of the cladistic method are entirely laudable and the method has been used in excellent taxonomic work. On the debit side, it is accompanied by a good deal of unnecessary jargon and has led to statements about evolution and about the assumed perfection of the method which are downright silly. It is not certain that a node in the cladogram should always connect three differently named taxa. Instead, it would sometimes be reasonable to think of the root form and one of the two branches as being the same. Sometimes the available evidence makes it reasonable to treat the material as if more than two branches could arise from a stem. In cladistics it is not possible to use a statement such as 'reptiles are ancestral to birds and mammals', because that takes two distinct groups and compares them with a group characterized by lack of the features which make them birds or mammals. While this may be formally correct, and the statement formally unacceptable, biologists do not usually misunderstand the sense in which such statements are made. In the end, the goodness of the classification depends on the choice of the characters to be used. No matter how pure his procedures, the cladist has to decide which features are important, and in this respect he does not differ from his classical counterpart. Experience and judgement, rather than the logical system, must be used to deal with the problems of the convergence and correlation.

In this discussion we have avoided attempting a neat definition of species. The concept comes from the general perception of patterns of organization coupled with the classification process. In common language, species are groups with permanent characteristics distinct from other groups. This is true of kingdoms and categories beneath them, such as classes and orders, but species come at the end of the list. Below them, in bisexual organisms, are individuals which interbreed and therefore undoubtedly fail the test by having common ancestry. The idea of species is therefore intimately tied to reproductive isolation, which allows species to remain distinct when living side by side. It is an old concept, implicit in the story of Noah's Ark. Biology recognizes the problems, and the exceptions they create, such as horizontal gene transfer between higher taxa, unisexual organisms, clonal reproduction, species introgression and evolution itself. Nevertheless, the most useful attribute of species in relation to classification is reproductive isolation (demonstrated or supposed) from other similar groups.

1.5 SUMMARY

The great variety of organisms became apparent in the eighteenth and nineteenth centuries, at a time when scientists concentrated on the essential features of each type, or taxon, which distinguished it from the next. Interest in the possible significance of the variety within species came later. Arguments against evolution and adaptation which were put forward were that we do not see intergradations between species and that many similar species often share the same locality. Twentieth century biology has been enormously successful in demonstrating the common molecular features which all organisms share. It has also confirmed the multiplicity of species and increased our estimates of their number by a factor of ten or more. Classification of organisms is based on overall similarity (phenetics) or on the identification of shared features (cladistics). Both methods confirm that species are natural units, almost always discontinuous with each other, each organism being an integrated system of interacting parts.

1.6 FURTHER READING

Mayr, E. (1970) *Populations, Species and Evolution.*
Mayr, E. (1976) *Evolution and the Diversity of Life.*
Ridley, M. (1993) *Evolution.*

Explanations 2

2.1 ADAPTATION

If there are some tens of millions of species each distinct from the next, then why do these groupings occur? Are they a necessary consequence of the way the world is structured? Is it possible to predict the number which will be found in any one part of the world? In order to attempt to answer these questions, it is necessary to have some idea of the way in which features of the organisms are related to the places in which they live. A characteristic enabling an organism to survive under particular environmental conditions is called an adaptation. The process by which such characteristics develop is also called adaptation, so that pattern and process are sometimes confused. Understanding of both is essential to unravel the relationships which develop.

It is useful to distinguish three categories of adaptation. The first of these consists of what may be termed special adaptations. This is a catch-all term for examples of the specific relationship between features of an organism and the conditions in which it lives. Examples such as the flippers of seals, the prehensile tail of the kinkajou or the fluked tail of the whale are special adaptations. They are noticeable because they fit the bearer for a way of life which is somewhat aberrant for the group to which they belong. Convergence of dissimilar organisms as a result of common environmental requirements is a problem in systematics resulting from special adaptation. There is no end to such examples, which may be physiological, biochemical or behavioural as well as morphological. They permit members of a species at worst to tolerate and at best to thrive in their environment and to compete successfully with members of their own and other species. The exceptions also imply that the typical condition for the group is an adaptation.

It sometimes seems that there is another category of adaptation, which may be called general. Whereas special adaptations manifestly match their bearers to their environment in particular ways, organisms sometimes exhibit traits which reduce their dependence on the environment, usually by increasing homeostasis. One of the most obvious examples is the endothermy of birds and mammals compared with ectothermy of other tetrapods. Mammals and birds expend a considerable amount of energy maintaining their body temperatures within narrow limits. As a result they can live successfully in conditions of extreme heat and cold which would be impossible for reptiles or amphibia. Another example is the large-yolked egg of birds, compared with the small-yolked eggs of amphibians, or the large-endosperm seed of many flowering plants. Others are internal fertilization, rather than external, and vivipary, rather than egg laying. In the past biologists have been accustomed to the idea of a rise of life from primitive and presumably primaeval forms to modern and advanced ones. From this standpoint it is easy to accept the idea of steadily improving general adaptation. Nowadays we are more likely to stress the modular relation between the organism and its environment; the tapeworm is just as well adapted as its mammalian host. It would be pointless for it to be capable of homeostasis since it lives in a buffered environment, and there is no sense in which one type of adaptation can be said to be more general, and perhaps better, than another. In any case, it is frequently difficult to decide which are the general adaptations. Bony fish have an efficient vascular system to circulate the blood. In the heart there is a single ventricle which pumps blood to the gills to receive oxygen. From there the blood passes to the brain and the visceral organs. In birds and mammals there are two ventricles, one pumping blood to the lungs for oxygenation and the other passing the oxygenated blood to the brain and body. The position of the arteries differs in the two groups but the systems they possess are probably equally efficient. In amphibia there is a single ventricle with an incompletely closed septum which allows some mixing of oxygenated and deoxygenated blood. A situation like this sometimes occurs in man as a result of disease or of congenital malformation; it is severely deleterious. Is the amphibian condition a stage in a process of general adaptation leading to the final two-chambered condition? Almost certainly not. Amphibia live in an environment where some respiration takes place through the lungs and some through the skin. They are inactive for long periods, then show sudden activity which results in a marked change in heart beat and oxygen need. The arrangement of the heart accommodates this variable pattern of breathing and is an adaptation to allow for it. It is likely that many, perhaps most, examples of possible general adaptation can be interpreted as specializations in a comparable way. Nevertheless, the modern view has probably gone too far in its relativism. When we survey the evidence of the fossil sequence it is undeniable that the living world has become richer with time. New forms have succeeded old, very often because they exploit opportunities previously present but unexplored. Cuvier's Radiata are marine, ectothermal, isotonic with their environment, externally fertilizing and archaic. His Vertebrata are more modern, and distributed over a much wider range of environmental conditions. Part of their success may be put down to the possession of better homeostatic systems than hitherto existed. It is certainly worth distinguishing the general from the special adaptation to draw attention to these changes with time, even when it is difficult to place a given example in one category or the other.

A third category of adaptation which may be distinguished for clarity is the developmental. This is any feature which arises in an individual organism as a direct response to the environment. In birds and mammals thickening of the skin where it receives wear falls into this category – examples include thickening on the sternum in chickens, the ischial callosities of baboons and the thickening on the fingers of people who write a lot with pens or pencils. The ability of the skin to bring about the change is an adaptation, but the change itself can also be spoken of as an adaptive response. Many examples may be found in relation to metabolism, where enzyme production is elicited by certain concentrations of the substrate or switched off by increasing concentration of the enzyme. Embryonic development depends on a host of adaptive feedback responses to channel the growing body into a particular form. This ability to provide a flexible response increases the independence of the organism from its environment. As before, it may be difficult to make a clear distinction between developmental and other forms of adaptation, but naming it draws attention to another facet of the process.

2.2 ADAPTIVE INTERRELATIONSHIPS

It is useful to illustrate the way in which different types of adaptation interlink. Of all mammals, human beings occupy the widest range of habitats in the world, from tropical forests and deserts to polar ice. Unlike most mammals, the human body is almost hairless, and in place of hair there is frequently a deposition of melanism in the epidermis which turns the skin dark brown. Although there are exceptions, there is a general association of degree of skin pigmentation with latitude, so that people inhabiting regions near the equator are darker in colour than those in more temperate regions. This correlation occurs among different unrelated segments of the human population. Pigmentation becomes more intense as one moves from Scandinavia to equatorial Africa, as it does if one were to follow a transect from Siberia to New Guinea. Dark skins in Melanesians and Africans often make them look superficially similar although they are distantly related. The correspondence with latitude is remarkable in Europe because over the past 4000 years migration of peoples in the region has been predominantly from east to west. Successive waves have come from western Asia and crossed the subcontinent to fan out northwards to Scandinavia and the British Isles or southwards to the Iberian Peninsula and into north Africa. Nevertheless, the gradient, or cline, of increasing pigmentation is from north to south. Now, ultraviolet light is harmful to the skin and the melanin provides a protective layer. When exposed to sunshine people with fair skins are much more liable to malignant melanomas than those with dark skins. The correspondence of pigmentation and latitude may be an adaptation protecting individuals from the increasing levels of ultraviolet radiation which are likely to be experienced as one proceeds south. Why then is everyone not dark skinned? That would add a safety factor even in environments where relatively low levels of sunshine occur. One answer to this kind of question is that every adaptation entails a cost. Melanin is a product of tyrosine metabolism by fairly energy-extravagant pathways, and the presence of any system which uses energy but is unnecessary results in a less efficient organism. This explanation is widely used in discussion of adaptation, but in the present case is probably wrong. Pale skin in regions of low solar radiation is probably advantageous in itself. Ultraviolet radiation may damage the skin, but in lower doses it acts to catalyse the production of vitamin D in the epidermis. From there the vitamin is carried to the lining of the intestine, where it is involved in absorption of calcium. A low degree of pigmentation may therefore be valuable to facilitate absorption of calcium for bone production, especially in times and places where there is a low level of sunshine. The process of acquiring a tan in summer and of losing it in winter in a northern climate assumes significance in this context as a developmental adaptation matching the protective and metabolic properties of the skin to changing seasonal levels of solar radiation.

Other factors involved in the story are the availability of sources of calcium in the food and of foods which aid its absorption. Among these, milk plays a very important part. It is not only rich in calcium but also in the carbohydrate lactose which, along with vitamin D, helps to get calcium into the body system. Most mammals have a significant amount of lactose in the milk. The newborn child has an enzyme, lactase, which breaks lactose into glucose and galactose. In the majority of humans this lasts until weaning, when production declines. The enzyme is produced to cope with the need to digest milk at this time of very rapid bone growth, dependent on calcium intake. Along with calcium, a source of vitamin D is also required. This is often derived from the action of solar radiation on the skin, but may also be obtained from marine fish oils. It is interesting to note that seals and walruses, which have a high fish diet, do without lactose in the milk and cannot digest it. The vitamin D intake from the food is sufficiently high to provide a milk which allows absorption of calcium without it. Calcium may also be obtained from certain vegetables such as cabbage, spinach and the like. It appears that northern Europeans have been short of all these sources, so that not only are they lightly pigmented but they have adopted the widespread habit of drinking milk into adulthood. Europe is one of the parts of the world where there is a high retention of lactase production into adulthood, a condition which is probably inherited as a dominant. Adult lactase sufficiency drops in frequency as one reaches the Mediterranean region. It is high among certain peoples of Africa, such as the Bedouin, the Fulani of northern Nigeria and pastoral tribes in East Africa. It is absent or at a low frequency everywhere else. Adults who are not lactase sufficient cannot digest milk, which is usually absent from their diet. Thus, although low skin pigmentation and adult lactase retention in Europeans are probably associated adaptations to life at low sunshine levels on a poor diet, different correlations occur elsewhere. Eskimos see little sunshine but can get their vitamin D from fish oils. Dark-skinned pastoralists in Africa continue to drink milk as adults and are lactase sufficient although living under conditions where vitamin D production in the epidermis is not a problem. Presumably in this case the lactase sufficiency has been selected for as a result of a cultural pattern which has developed for other reasons. There are two other ways in which human beings can cope with digestion of milk. The gut bacteria, including the famous laboratory organism *Escherichia coli,* have to adapt to the dietary patterns of their host and can break down lactose into its constituent sugars. The genes for doing so have been much used in *E. coli* as a model system for studying gene action. Milk drinking by a lactase-deficient adult brings on unpleasant symptoms of indigestion, but these may decline somewhat with time as a result of adaptation by the gut microorganisms. This is a symbiotic interaction; the well-being of the host is improved and the chance of survival of the bacteria increased. Many such interactions between microorganisms and their hosts result in adaptations of a more formal status, such as the provisions for cellulose-digesting protozoa in termites and cattle or the musk-producing signalling organs of skunks, weasels and deer. Accommodation to lactose never reaches a similar level in lactase-deficient people, but an alternative, if one wants to use milk as a food for adults, is to allow part of the digestion process to take place before the food is eaten. This is what happens in the production of yoghurt and cheese. Fermentation has made milk products available as calcium-rich but digestible food to vast numbers of people intolerent to milk, including the Mongol peoples of central Asia. Absorption of calcium is reduced, but the products are available and in some parts of the world extremely important sources of food.

Another interesting example of a complex of adaptations concerns the horns and antlers of various herbivorous mammals. All the examples are artiodactyls, which are primitive browsers, feeding on the foliage of trees. They include the giraffes, of which modern examples have bony projections on the head but some fossil forms possessed branching antlers. In the Bovidae, the cattle, sheep and goats have permanent horns usually present in both sexes while antelopes have horns in one or both sexes. The Cervidae or deer possess antlers which are deciduous and usually present only in males. These animals are frequently polygamous. Males compete for females and the appendages on the head form an important part of the weaponry used in competition. Fighting is usually ritualized. Stags lock antlers and push each other to a state of exhaustion; big-horn sheep run at each other and have reinforced fronts to their skulls to take the impact. The objective is to dispossess or discourage the opponent, rarely to kill it. Darwin pointed out that stags with short pointed horns are often more dangerous than ones with full branching antlers, yet the large-antlered individuals often win the hinds. The antlers or horns are signalling adaptations which advertise the status of their bearers, as well as weapons of attack or defence. Several biologists have suggested at different times that they may have other functions as well.

Horns consist of bony extensions of the skull, the core, enclosed in a keratin cover. Between these two structures there is heavily vascularized tissue. Arteries extend up the horn and ramify extensively, one artery joining another. The blood pumped into the horn is collected by veins which drain to a large sinus behind the eye. These veins may be dilated or constricted and the whole complex structure is designed so as to have a temperature regulatory function. If one touches the horn of a goat it often appears very warm, and experiments have shown that the amount of vasodilatation is related to the amount of work the animal does, so that the horn dissipates metabolic heat. It has been calculated that in the Asiatic buffalo something like 15% of the daily metabolic heat is lost through the horns. So horns are weapons, signalling devices and also thermostats – what about antlers?

The antler of an animal such as a red deer is a bony extension growing from a base, or pedicel, on the skull. It is produced annually, reaching its full extent and then dropping off in 6–8 months. During growth the antler is made of spongy bone and is covered with the so-called velvet. This is a very vascularized and glandular tissue covered with fine hair. Arteries pass up the antler and anastomose directly with veins rather than passing through a capillary system. The velvet bruises easily, bleeds easily and in cold weather suffers badly from frostbite. It is therefore a bad protective covering but very well designed for temperature regulation. Bernard Stonehouse has suggested that one of the reasons for having deciduous antlers is that a seasonal temperature regulatory device is needed by males of many species of deer. Antler and body growth occur in spring when females are giving birth to and feeding young. At this time the nutritive value of the food increases, and must be great enough to support the reproductive outlay of the females. Rutting occurs in autumn, by which time the bone of the antlers has become dense and the velvet has been lost.

Most deer have this pattern of life. Tropical species are exceptions, living in a more constant environment. They tend to be small and to have no antlers, sometimes fighting with enlarged canines instead. Reindeer have antlers in both sexes and in the calves, the reason for which is unknown; possibly the antlers are used in foraging for food. Flattening, branching and coiling of antlers or horns increase the surface area while keeping the projections within a reasonable distance of the head. If these structures have a thermoregulatory function, a correspondence between their surface area and body weight would be expected and is sometimes seen. Within species, individuals of different sizes show a greater rate of increase in size of the horn or antler than of body size. In tropical African savannah regions the antelope share their feeding areas with hornless zebra. If one type of animal requires horns to regulate temperature, how is it that the other gets on without them? The answer could be that the zebra is differently organized and has some other kind of thermoregulatory adaptation. Despite the fact that they live in the same places, however, the two types of animal eat different food. Antelopes are browsers, plucking the highly nutritious leaves from plants with their lower teeth and the horny plate which replaces the upper incisors. Zebras, like other horses, possess a good set of teeth in the upper and lower jaws, which they use to bite off the stems, so that they achieve a lower energy intake for a comparable amount of work. It may be that the zebra does not experience the same problem of excess metabolic heat. Finally, to conclude this survey of reasons why some mammals have extensions on their heads, it has been suggested that they play a part in distant, as well as close-up, signalling. It is very obvious whether a deer or antelope is looking towards or at right angles to the observer. This information is important to the stalker and as A. Zahavi has suggested, presumably also to conspecifics when establishing and defending their territory. Any extension above the head, no matter how small, could serve as an indicator of the viewing direction, providing a reason why the process started, before a defensive or thermoregulatory function was possible.

The examples of vitamin D in man and horns and antlers in ungulates are not presented simply as facts. Much of the argument is speculative and the data incomplete. Writers could be found who would contest many of the statements made. What they do illustrate, however, is the way in which adaptations may be expected to interrelate to one another. If we start at any point in such a study we may expect it to lead us eventually through all the systems of the organism, because living things are integrated wholes. Although we examine animals and plants a bit at a time, life, death and relative success operate at the level of the complete individual. This could perhaps be called the Cuvier principle, since recognition of interrelationship was the basis of his natural classification.

2.3 TROPHIC SYSTEMS

Plants and animals are distributed in space. One of the ways of studying diversity is to examine a particular region or locality to see what sorts of association occur. A mixture will be found which not only consists of many species but exhibits a pattern depending on the way in which energy is obtained. Energy is derived from the sun, but only photosynthetic organisms, the primary producers, can fix it. All other kinds of organism depend on exploiting this primary source. Tracing the paths by which energy disperses through the system provides a start to understanding the causes of diversity. The interactions of the food chain impose what is called a trophic structure on the community of organisms. One way of describing the pattern is to record the number of individuals belonging to each step in the chain. The examples shown in Table 2.1 come from classic work on trophic systems summarized by Odum (1959). Table 2.1(a) gives numbers of individuals in a field of grass. There are vastly more plants than there are herbivorous invertebrates which feed on them, and many more of these than of carnivorous invertebrates and vertebrates. The consumers are labelled Cl to C3, but actually C3 is dependent on C1 rather than on C2. Numbers of carnivores specializing in eating other carnivores are rarer still. In this situation they could only consist of a few hyperparasitoid insects. The list is incomplete because it ignores decomposers; bacteria, fungi, worms and any other organisms which utilize the available dead organic material. In one respect it is misleading too, because individuals at different levels are very different in size. No information is provided on the biomass present in each layer, and biomass gives a clue to the reason why such trophic chains are short. Another problem with the individual as a unit is that plants are often present as extensive clones, which are not directly comparable to single animals. In Table 2.1(b) the pyramid for a freshwater community is given in different terms. The left-hand column shows biomass for the same four levels and for the decomposers. The attenuation as one proceeds away from the producer level is less extreme, but it confirms the small amount of living matter which can be supported by the energy tied up in photosynthetic organisms. Biomass as a unit indicates availability at a given moment. The next column in Table 2.1(b) provides information on the dynamics of the system, since it is expressed in kilo-calories per square metre per year. This shows the total energy flow, part of which is fixed as biomass. In the next column we have the fraction of energy transferred from one level to the next. Only 2% or so of the available solar energy is fixed, depending on how the calculation is done. Efficiencies of transfer are higher at other levels, but still do not rise above 20%. These figures represent the energy used to drive the system as well as to maintain a standing mass of organic matter. The final column shows the ratios of biomass. The difference

Table 2.1 Ratios of individuals and of energy in different trophic levels

(a) Number of individuals at different trophic levels in a field of grass:

C3	birds and moles	3
C2	spiders, ants, predatory beetles, etc.	354 904
C1	herbivorous invertebrates	708 624
P	producers–green plants	5 842 424

(b) Production of different trophic levels at a freshwater site at Silver Springs, Florida:

		Biomass ($g\ m^{-2}$)	Energy ($kcal\ m^{-2}yr^{-1}$)	Ratio to energy (%)	Ratio to biomass (%)
C3	top carnivores	1.5	21	6	9
C2	other carnivores	11	383	11	5
C1	herbivores	37	3368	16	17
P	producers	809	20 810	2	—
D	decomposers	5	5060	21	4

Data from several sources summarized in Odum (1959).

between the columns is explained partly by differences in individual size. Decomposers, being mostly small, have high metabolic rates and low biomass for the energy used, whereas larger animals are more efficient in the sense of tying up the energy as living matter.

These examples show us that the primary structure of communities is imposed by the trophic system. Each level depends on a different level for its energy income. Some of the species diversity can be related to ways of exploiting these resources. Thus, land plants typically have leaves in which photosynthesis takes place, and in order to support these they have woody stems and branches. Herbivores therefore have two types of structure available, the wood and the leaves, from which they may obtain their energy. In principle too, an animal may have a single role in the system, as a carnivore, a herbivore or a decomposer or it may obtain part of its food in each of these ways. The existence of the trophic system itself, however, does not do much to explain the multiplicity of species.

2.4 HABITAT DISTRIBUTION

The world is composed of a great variety of habitat types. These may be terrestrial or aquatic, freshwater or marine. Terrestrial habitats are affected by geology and topography, climate and weather. It seems obvious that we should not get the same associations of organisms in a terrestrial location such as the field referred to in Table 2.1(a) as in the shallow water of Table 2.1(b). Moles are associated with the field and phytoplankton with the water, and they have all sorts of characteristics which indicate to us that this should be so, not least their methods of respiration and energy gathering.

There are, however, terrestrial algae and aquatic insectivores, and groups such as annelid worms are not obviously better equipped to live in one type of habitat than the other. In Britain, the earthworm *Allolobophora chlorotica* is found in pastures, but also under logs, in ditches and even in the bottom of lakes. Species diversity follows habitat diversity. The extent to which it should do so is not self-evident, however, but depends on adaptive constraint. Every adaptation imposes a cost, metabolically, and within an organism all aspects of adaptation interact. An organism may be highly adapted in some respect, but as a result suffers a limitation in some other respect. If we could quantify these interactions more exactly it would be possible to predict whether a given amount of change in environment should require replacement of one species by another, or whether a single species could cope with both sets of conditions. As it is, we have to make do with general rules derived from observation. Geographical transitions from one environment to another are accompanied by changes in species composition. It is inevitable that this should be so, because species are adapted to particular conditions. As conditions change they become less well matched to them and inviable, or if not inviable, less competitive with more critically adapted species. A good deal of species diversity can certainly be put down to environmental or habitat diversity, although how much is unclear. There are probably many more species than the minimum set by considerations of adaptive constraint.

Assuming some level of constraint we can see how quite a lot of diversity is generated, following the 'woody plant' analogy used above. A given area has some dry patches of soil and some damper patches. Adaptive considerations dictate that each should be colonized by a different type of plant. Because they are different, the leaves are favoured by different types of herbivore. Competition between the plants results in one having woody stems while the other has not. A place is made available for a third type of herbivore, and a different kind of specialist decomposer is required. With three herbivores to choose from as food, there may be room for two types of carnivores. This model-building game suggests that species number should increase at a faster rate than habitat diversity. Habitat diversity must increase with area, possibly in direct proportion to it. Species number is therefore more likely to increase logarithmically than linearly with area. Similar considerations indicate that if we could take a random sample of habitats and count the species in them then the number of species should be log normally, rather than normally, distributed. Part of our surprise at the number of species is due to a habit of assuming arithmetical distributions; a particularly diverse habitat will have two or three times as many species as a less diverse one, not an extra two or three. If analogy is possible, it might be with the subjective effect of speed. If we double the speed at which we drive a car there are sometimes surprising consequences. Physically, the reason is that $E = mc^2$, but subjectively we expect the forces imposed to change linearly with velocity. The physical system is consistent but we have to get into the habit of recognizing this fact.

2.5 NICHES

Study of the way species live together in communities has led to the definition of the niche. Whereas the habitat of an organism is the place in which it lives, its niche is what it does and how it does it. Empirical observation shows that closely related organisms often live in different niches within the same habitat. It is not simply the case that adaptation to one of these sets of conditions renders a species seriously maladapted to the other. In the absence of one of them, the other species would be at home in the environment of both. When both are present, some fine partitioning of the available resources takes place between them. The naturalist's view of communities abounds in examples of trios or quartets of similar species occupying very similar conditions. Sometimes they are so similar as to be almost indistinguishable. Although some earthworms are surface feeders and others feed only in the soil, finer distinctions also exist. *Allolobophora chlorotica*, mentioned earlier, and *A. viresceus* are two species whose separate status has been repeatedly questioned, and confirmed only by breeding and isoenzyme analysis. The same means was used to demonstrate that the striped and unstriped worms of compost heaps really are the separate species *Eisenia fetida* and *E. andrei*. In other respects the differences between them are minute, yet they commonly coexist. On northern European shores there are two species of the 'flat periwinkle', *Littorina obtusata* and *L. mariae*, and four species of the 'rough periwinkle', *L. saxatilis*, *L. nigrolineata*, *L. neglecta* and *L. arcana*. There are small differences in ecology and life history between the members of each group, but generations of professional zoologists used them as the subjects of biological field courses year after year before the differences were sorted out. In British hedgerows and meadows, sets of similar species of butterflies are to be found. There are three species of 'whites' Pieridae, which feed on crucifers, three species of grass-feeding 'browns', Satyridae, three species of 'skippers', Hesperidae, two species of 'blues', Lycaenidae, two or three nymphalines, all eating nettle, and so on. Groups of this kind occupy the same habitat but distinct niches. The same pattern is seen when we examine the common birds of woodlands. In deciduous woodland of Britain, there are three species of tits, Paridae, the blue tit, the marsh tit and the great tit. They often form mixed species flocks in winter which travel around together feeding. When nesting in broadleaf woodland they tend to partition the environment into different feeding levels when searching for insects, the blue tit at the tops of the trees, the marsh tit below and the great tit near the ground. In the coniferous woodland of Scandinavia, a

Table 2.2 Birds of the genus *Parus* from North America and Europe, showing their main habitat, size and shape. There are morphological differences associated with habitat between species in the same region and parallels between the two regions

North America	Main habitat	Weight (g)	Beak length and depth	Europe
Tufted tit *P. bicolor*	Broadleaf	22	Long, broad	Great tit *P. major*
Bridled tit *P. wollweberi*	Broadleaf/ pine	10	Short, broad	Blue tit *P. caeruleus*
Black-capped chickadee *P. atricapillus*	Broadleaf and open conifers	12	Medium, medium	Willow tit *P. montanus*
Carolina chickadee *P. carolinensis*	Broadleaf	10	Medium, broad	Marsh tit *P. palustris*
Mountain chickadee *P. gamebeli*	Montane pine	11–12	Long, thin	Crested tit *P. cristatus*
Chestnut-backed chickadee *P. rufescens*	Spruce	10	Long, thin	Coal tit *P. ater*

Adapted from Lack (1971).

different habitat, their place is taken by the coal tit, the crested tit and the willow tit. These also separate their feeding niches by height. In North America, species in one of the subgenera are known as chickadees. There, broadleaf and coniferous woodland are occupied by a different set of species, between which evidence of a parallel set of niche specializations is apparent. Comparable species from the two continents are shown in Table 2.2, which also shows that the size and shape of the beak vary in relation to the particular feeding stations of the species.

Evidence for niche specialization and the concept of the niche have been the subject of much study. Gause's principle states that no two species can coexist if they occupy the same niche. This is effectively a definition of the niche. Minimum adaptive differences are apparent between species which do coexist. In insectivorous and seed-eating birds it is found that a pair of species may be the same size when they live apart, but if they coexist they differ in beak dimensions by a minimum of about 15%. There is said to be character divergence. The difference ensures that there is a maximum overlap in food resources which is not exceeded. This rule is supported by the evidence in Table 2.2. Species abundance is partly due to the existence of many similar species in distinct but similar niches within a habitat.

2.6 ADAPTIVE RADIATION

When habitats in isolated regions are filled with distinct communities of animals and plants these groupings are termed adaptive radiations. For terrestrial organisms separation of continents and islands results in a sort of natural experiment where we can see what type of community develops. Very often striking parallels occur in different places. On a grand scale, the marsupial fauna of Australia contains many species with similar adaptations to placental mammalian counterparts. Marsupial moles and anteaters are extraordinarily similar to moles of the order Insectivora and the South American anteater. The koala resembles some lemurs in habit and food, the cuscus resembles a tree porcupine and the Tasmanian wolf a placental wolf or dog. Sometimes the morphological resemblance is less exact but the pattern of life exhibited is close. The large kangaroos of Australia are mobile herbivores exploiting hot dry grassland, like African antelopes. Like them, too, males fight for females, using their feet instead of horns in these contests. In both cases there is considerable sexual dimorphism. Although superficially different, the two groups show many similarities in habit of life. Parallels can also be seen among the extinct horse-like and tiger-like marsupials of South America. Within the placental mammals, the pattern is repeated. In South American forest, hystricomorph rodents such as the capybara, the paca and the agouti occupy the niches filled in Africa by hippopotamus, chevrotain and small deer such as duiker. The African pangolin resembles very closely the South American armadillo. The Old World tits and the New World chickadees represent parallel radiations within the family Paridae. Land snails tend to be either high spired or disc shaped, with few species of intermediate shape. The two modal shapes probably define particular kinds of niche. Among the terrestrial snails studied by Wollaston on the Madeiran islands, disc-shaped Palaearctic groups such as the Zonitidae are missing and Helicidae appear to fill some of the niches thus left vacant. The high-spired families Enidae and Chondrinidae are absent and their place is taken by species in the Ferussaciidae. Adaptive radiations show us that certain kinds of community are possible, with major types of specialization and similar groups of closely related species in similar niches. In the past, when new territories awaited discovery, the explorer hoped for, and sometimes discovered, something totally new but experience allowed the general pattern of life to be anticipated. Nobody could predict the platypus, but given experience of the great auk a seaman from northern waters would be unsurprised at encountering a penguin for the first time.

2.7 CONCLUSION

The evidence suggests that a particular set of physical conditions in a given part of the world, that is to say, a habitat within a region, generates a particular set of niches which will be filled by analogous sets of species of plants and animals. In this view, the niche seems to be a real entity. If it can be identified in the Mediterranean and is occupied by a particular type of animal we can anticipate a similar association in other areas with Mediterranean climates, in Australia or Chile, for example. The organisms involved are taxonomically distinct but similar in role. We have got a little way towards explaining the existence of millions of species. As a first step, it can be said that trophic systems require a minimum level of diversity. If we add to that the enormous habitat diversity which exists on Earth we are doing much better. Species number and habitat diversity are likely to be logarithmically related, so that geographical diversity implies the existence of many species. But as many as tens of millions? The difficulty in answering this question arises from two sources. On the one hand, it is impossible to predict the relationship between species number and habitat heterogeneity unless one knows the average breadth of the ecological conditions to which a species can become adapted. So little is understood on this subject that it is difficult to know where to begin. In addition, we have to take a longer look at the concept of niche. If niches can really be said to exist, then species may be expected to fill them. Suppose, however, that when similar species accidentally come into contact they adapt to minimize the extent to which they interact. In that case, the number of species is to some extent a function of the number of such contacts which have been made. Niche diversity is then a consequence of, rather than a prerequisite for, species diversity.

2.8 SUMMARY

Adaptation is the condition of being fitted to survive in a given environment, and an adaptation is a characteristic enabling an organism to survive. These terms are used in a relative sense. In given conditions, some individuals are better adapted than others and there are no circumstances in which an individual can be said to be perfectly adapted. It is useful to distinguish special adaptations, which equip the individual to cope with particular conditions, from general adaptations, which increase the range of conditions that can be tolerated. A third category, developmental adaptations, consists of characteristics which are elicited by environmental effects during the life of an individual. Since organisms are integrated systems, improved adaptation in one respect often results in reduced adaptation in another. The patterns of assemblage of organisms into communities in a given area may be described under four headings.

1. The first is the trophic or food-chain organization of species according to the way they obtain their energy. All organisms are photosynthetic, or decomposers of organic material or eat other organisms. Food chains tend to have not more than five or six links. Energy input comes from photosynthesis and much energy is lost in transfer from one link to another. This is probably the main reason for the shortness of chains.
2. The world is made up of a variety of different physical habitats, and each habitat has species adapted to live in it. The number of distinguishable habitats is to some extent determined by the adaptive tolerance of the species.
3. Within habitats, niches may be recognized. A niche is the array of conditions utilized by a given species, in which it is competitively superior to other species. There may be niche overlap in some measurable features, but no two species may occupy identical niches. The niche is therefore a reflection of the species which coexist and may come into competition.
4. It is a general observation that isolated areas of land or bodies of water often have their own communities, which parallel each other but consist of taxonomically distinct species. These assemblages are known as adaptive radiations.

2.9 FURTHER READING

Futuyma, D.J. (1986) *Evolutionary Biology*.
Gaston, K. and Spicer, J.I. (1998) *Biodiversity. An Introduction*.
Krebs, C.J. (1994) *Ecology*.
Mayr, E. (1970) *Populations, Species and Evolution*.
Ricklefs, R.E. (1997) *The Economy of Nature: a Textbook of Basic Ecology*.
Ridley, M. (1993) *Evolution*.
Wilson, E.O. (ed.) (1988) *Biodiversity*.

Dynamics 3

3.1 EVOLUTION AND NATURAL SELECTION

So far, the question of species diversity has been considered in static terms. Careful observation allows us to determine species. These are to be found in certain types of habitat, and different assemblages are found in different parts of the world. Explanations consist of establishing laws which define the relation of species abundance to ecological conditions. This approach would have been familiar to biologists in the early nineteenth century. The evidence was becoming available then as a result of two factors. One was an increasing willingness to go out and learn by observation, to pursue natural history instead of relying upon philosophical discussion. The other was the vast increase in information about geography and nature which accrued from aggressive and enquiring exploration and colonialism. Francis Galton, a pioneer in the study of biometrical inheritance, had a background typical of many British biologists at the time. He was rich, energetic, intent on finding explanations for natural phenomena and a member of a group of like-minded relatives and acquaintances. He was, writes an editor of his *Narrative of an Explorer in Tropical South Africa* (1853), 'the third son of Samuel Tertius Galton, a banker in Birmingham, in whose family the love of statistical accuracy was very remarkable, and of Violetta, eldest daughter of the celebrated Dr Erasmus Darwin, author of *Zoonomia, The Botanic Garden*, etc.' In explaining his travels he states, 'The motive which principally induced me to undertake this journey was the love of adventure. I am extremely fond of shooting, and that was an additional object; and lastly, such immense regions of Africa lie utterly unknown that I could not but feel that there was every probability of much being discovered there, which, besides being new, would also be useful and interesting'. What an ideal position from which to try to make sense of the patterns one sees about one! The key was to discover the dynamics, to show that the patterns must come about. This was an obsession with scientists in the mid-nineteenth century, just as the code of life became the central obsession of biology in the mid-twentieth century.

The way the evidence pointed to a solution can be seen by examining the experience of some of the principal participants in the quest in Britain. Charles Lyell was born in 1797 and died in 1875. Private means allowed him to become a geologist and palaeontologist. He went to Madeira with Joseph Hooker, and between them they noted not only the endemic fauna and flora but the existence of fossil corals and plants in the rocks many hundreds of feet above the modern sea-level. This evidence contributed to formulation of the principle of uniformitarianism, that events in the geological past have followed similar patterns to those seen today. Joseph Hooker (1817–1911) was a botanist. He became director of Kew Gardens, travelled to the Antarctic with James Ross and also to Sikkim and other parts of the Himalayas. He was author of a *Flora of Tasmania*, a study of an adaptive radiation in an isolated region.

Alfred Russel Wallace was another long-lived Victorian (1823–1913). He came face to face with exuberant biological diversity in the Amazons, where he went with his brother William and H.W. Bates. Unlike most of the others, this trio was not rich, but earned their livings while pursuing their interests in biology by collecting birds and insects for collectors in England. William Wallace died in South America in 1848; Bates stayed there for some years before returning to England to become assistant secretary of the Royal Geographical Society. Alfred Wallace continued his collecting in the islands of Indonesia, where he noted the faunal transitions which occurred between Bali and Lombok and between Borneo and the Celebes. While detained on the island of Ternate through illness, he formulated his theory of the driving force producing the patterns.

Charles Darwin was born in 1809 and died in 1882. He was grandson of Erasmus Darwin. As a young man he travelled around the world on the survey ship *Beagle*, taking with him and what is more, reading, Lyell's *Principles of Geology*. He experienced the diversity of tropical flora and fauna in Brazil, and studied an extraordinary radiation of fossil marsupials in Patagonia. He was introduced to endemic variation on the Galapagos, where each island has its own distinct species of giant tortoise. As is well known, he worked slowly towards his conclusions about evolution and its causes on his return to England, only going into print after receiving a draft of the same ideas from Wallace. Their joint communications were published by the Linnean Society in 1858.

The collected evidence indicated that diversity was apparently associated with isolation. Competition and diversity are negatively correlated. Put together with two other pieces of evidence, this insight was developed into a dynamic theory of evolution by natural selection. The necessity for evolution arises from two sources.

1. All life is derived from other life. Following Aristotle, separate creation and continuity from one generation to another had been accepted for higher forms of life, but spontaneous generation for lesser ones. The idea of spontaneous generation died out in the nineteenth century, partly as a result of increasing awareness of the complexity even of the simplest organisms, but more particularly because of experimental demonstration of the effect of sterilization.
2. There is a fossil sequence. Geological horizons are convincingly arranged in a time series. The rocks contain extinct forms but lack modern ones. Furthermore, organisms with lower levels of general adaptation tend to predate those with higher levels in their time of first appearance.

It follows that the changes and the variety must be the result of evolution. If not, then the first or the second proposition, or both, must be false. This caused problems for those who had religious reasons for discounting evolution. The marine biologist P.H. Gosse suggested that the Creator may have arranged the fossils so that we experience a sensation of continuity. T.V. Wollaston claimed a scientific basis for his rejection of evolution. He was not fully cognizant of the fossil evidence, however, and proposed a catastrophic explanation, the drowning of lost Atlantis, to explain the richness of the faunas he studied. In fact the geological evidence was quite fragmentary when it was first used in the cause of evolution; subsequent work has overwhelmingly strengthened the conclusion. Since evolution occurs, taxonomic similarity can be equated with phylogenetic similarity. There must be continuity between different forms of life, one type being descended from another, despite the apparent lack of continuity now.

The contribution of Darwin and Wallace was to provide the mechanisms to drive the process of evolution. This is based on four lines of evidence and three conclusions, as follows.

1. Organisms have a capacity for increase in numbers, which is in principle exponential.
2. Despite this capacity their numbers remain, on average, constant. Consequently, struggle for existence must occur.
3. Within species, individuals vary in their characteristics. Consequently, the competition which has been demonstrated must lead to differential survival of those best fitted to survive. A process of natural selection must operate.
4. Some of the variation between individuals is inherited. Consequently, natural selection gives rise to evolution.

Through differential survival, natural selection may modify the properties of a population within a single generation. Because of the inheritance of many of the traits concerned, it leads to special and general adaptation in a changing world, following the vagaries of the environment. It may result in increasing general adaptation in a constant environment, because any fitter form which arises tends to increase in frequency to replace less fit relatives.

The full title of Darwin's most influential book, first published in 1859, is *The Origin of Species by Means of Natural Selection or the Preservation of Favoured Races in the Struggle for Life*. The more thought one gives to this title, the more thought provoking it becomes. The innovation of Darwin's approach was the identification of natural variation, coupled with selection, as the mechanism for adaptation. We have a picture, developed in the previous sections, of species associated with one another in more or less congruent groups and fashioned so as to fit their environments. Yet species do not usually show evidence of transition from one to the next, adaptation and speciation are not the same thing, and so it is not self-evident that natural selection generates new species. It is necessary to look more closely at the concept of species.

3.2 DEFINITION OF SPECIES

Taxa are categorized by a system of names, of which the most important is the Linnean binomial attached to species. *Cepaea nemoralis* is a species of snail in the genus *Cepaea*. It belongs to the subfamily Helicinae of the family Helicidae, like its close relative *C. hortensis* and the larger *Helix aspersa*. A somewhat different snail *Arianta arbustorum* lives in similar habitats to these three. It also belongs to the Helicidae but is classified in another subfamily, the Ariantinae. Systematists have differed on how closely related these taxa are, but there has never been any doubt that the four species are distinct. Different criteria for making such distinctions have, however, been used at different times. In his book *Animal Species and Evolution*, first published in 1963, the eminent American evolutionary biologist Ernst Mayr has distinguished three concepts used when defining species.

1. The typological species concept. This is the morphological species of the museum taxonomist. It is recognized by its common features and distinguishable from other species because of the natural discontinuities so frequently observed. The process of defining the species is no different from that used at other taxonomic levels, for the subspecies, genus, family, etc.
2. The non-dimensional species concept. This is the concept used by the field worker studying associations in the same place at the same time. The species level corresponds to what we have described as niches. There is only one species per niche. It may usually be distinguished

morphologically from similar species, but is also recognized by ecological and behavioural differences. Naturalists become very adept at recognizing the true distinctness of extremely similar forms. The ideas of the non-dimensional species and the typological species are not very different, but whereas the museum taxonomist takes what comes and classifies it using morphological characters for the most part and a preferred taxonomic procedure, the field naturalist is dealing with species whose distinctness has been sharpened by their adaptation to different facets of the environment. His task is made easier to that extent.
3. The biological species concept. This is the modern concept of species, and rests upon the fact that the species, like groups above them, are reproductively isolated, while individuals within species can interbreed.

Mayr defined species as 'groups of actually or potentially interbreeding natural populations which are reproductively isolated from other such populations'. On this basis, species are defined by their distinctness more than by their intrinsic properties, and by their constitution as groups of populations, rather than as groups of individuals. The biological species is the fundamental taxon, because adaptation takes place within species, while species proliferation is a process intimately tied up with evolution which may or may not be accounted for by explanations of the process of adaptation.

The biological species is often very difficult to define in practice. It is usually impossible to test for interbreeding. When this can be done one sometimes finds that groups of organisms, which from all other considerations should be regarded as species, will in fact interbreed. The criterion cannot be applied to groups which usually reproduce asexually. Common plants, such as the brambles (*Rubus fruticosus*), present great difficulty because of the variety of forms which exist, which may or may not be entitled to the status of species. Bacteria such as *E. coli* present similar problems. They reproduce asexually most of the time with occasional sexual reproduction. Even so, genetically distinct clones may be isolated from the gut of a single human host, which are closely similar to clones from a different host, living hundreds of miles away. This suggests that even in bacteria biological species exist. The interbreeding criterion cannot, of course, be applied to extinct species.

Most biological species consist of groups of geographically separated subspecies, or races. These are often evidently adapted to local conditions. They interbreed freely but remain distinct because the zones of contact between the different subspecies are small. Such a grouping is known as a polytypic species, and is reproductively isolated from its most similar relatives, usually belonging to another polytypic species. Subspecies are always geographically separated, whereas species may overlap in range.

Ring species raise a problem of species definition but at the same time provide an indication of the importance of the biological species. The best-known example is the herring gull – lesser black-backed gull complex. In Britain, the lesser black-backed gull, *Larus fuscus*, and the herring gull, *L. argentatus*, appear to be distinct species with different appearance, and somewhat different habitats and behaviour. They are distinguishable by the colour of the wings, the eye ring and mandible. *Larus fuscus* typically nests on flatter coastal plains, *L. argentatus* on rocky bluffs and cliffs. When feeding at sea, *L. fuscus* is more likely to plunge dive, and is less likely to scavenge than *L. argentatus*. As one follows *L. fuscus* eastwards, however, through Siberia to Alaska a succession of different subspecies is found which change in appearance and character towards that of *L. argentatus*. These extend across North America to the eastern seaboard, and across the Atlantic. Sexual isolation operates where the forms overlap but genetic continuity probably exists around the ring. Such examples are rare, but significant because they are exceptions to the rule that discontinuities exist between species.

Like most concepts in this book, Mayr's definition of the biological species has been with us for some time. It is not without its critics. The difficulties emphasized are usually among those listed above. An alternative approach is to define the species by its genetic composition, as a group with few individuals genetically intermediate between it and its most similar group. A spirited advocacy of this point of view is to be found in Mallet (1995). This seems to us to be no more than the realization in genetical terms of the discontinuities discussed in Chapter 1. Given that species sometimes contain subspecies (geographically separated populations which are recognizably distinct in some features, but interbreed), the biological definition must always be imperfect. Evolution itself, and the ways it can occur, necessarily renders it so. Nevertheless, there is an essential transition point in a continuity of groupings. In bisexual organisms, individuals within species can undoubtedly interbreed, while interbreeding does not occur between genera or higher groupings. The species is intimately bound up with the shift from the one level to the other, and for all its difficulties, the biological species concept most nearly recognizes the essential features of the transition point. Reproductively isolated taxa can coexist, specialize and evolve in response to each other. When there is full genetic continuity these processes become population genetics. The biological species is therefore essential for a particular level of evolution. These considerations are taken up in Chapters 16 and 17.

3.3 CONCLUSION

In the foregoing pages it has been suggested that the number of species we find around us is a matter for surprise, needing explanation. It is argued that the millions of species are real entities. In taxonomy they comprise the fundamental level of interbreeding but reproductively isolated groups. Part of the answer to why there are so many is that the world is physically heterogeneous; different species function best in different parts of it and there is probably a logarithmic relation between species number and environmental heterogeneity. But that only takes us so far. The matching of organisms to their environments involves the idea of adaptation and of adaptive constraints. One species cannot occupy two very different habitats because adaptation to one of them interferes with adaptation to the other. Only if we understand the process of interference can we say how many species a given mosaic environment should contain. It is argued that interest in adaptation and its limitations, and in the patterns of species diversity, led nineteenth century biologists to the view that evolution occurs and that adaptation is the result of natural selection acting on inherited variation. It is therefore necessary to know how much inherited variation there is within populations of a species, how it arises and how it is maintained. These are the subjects considered in the next few chapters.

Wollaston raised two objections to the idea that adaptation results in transmutation of species. One was that we do not see transitions from one to another. Similar species are often compared in the same locality or habitat, however, and in those circumstances they are likely to be adapted to occupy distinct niches, so that they require different sets of resources and do not interfere with each other. These are not conditions under which transitions would be expected. When geographical distributions are considered the distinctions are less clear, and it may be difficult to distinguish contiguous species from contiguous races within the same species. Some of these transitions are probably examples of species formation. The example of ring species is important because it demonstrates the existence of reproductive isolation between parts of a taxon in which there is otherwise interbreeding. If we do not often see evidence of speciation, that may be because on average species persist for relatively long times. Palaeontologists classify morphological types as distinct species if they differ by amounts which would indicate specific difference in modern forms. The rate at which fossil species replace each other suggests that their duration is somewhere from one hundred thousand to one or two million years. If that is so, we would be lucky to alight on a speciation event by chance, and could only expect to do so under conditions particularly favourable to species formation.

The second difficulty seen by Wollaston was that there are many cases of groups of similar species apparently adapted to the same conditions but coexisting. If adaptation were the key, he argued, then there should only be one species. The existence of these groups of similar species, which are commonly observed, is not explained by what has been discussed so far. Niche specialization concerns how they coexist, not how they arise. An explanation for this replication phenomenon is the final stage in accounting for species abundance. It is considered in the concluding chapter.

3.4 SUMMARY

Understanding the causes of species diversity depends on the evidence that the species have evolved. We assume that, since the earliest living systems appeared, all subsequent forms have been derived from other living forms. In that case, the fossil record demonstrates that evolution has taken place. The Darwinian theory of evolution by natural selection provides a mechanism for the process. No other theory has the power to explain adaptation.

The theory is based on the observation that not all individuals in a population are the same. Those best adapted to a given set of conditions have a greater potential for survival and increase than less well-adapted individuals. At least some of their characteristics are inherited, so that from one generation to the next these well-adapted features increase in frequency.

In sexually reproducing organisms, species are groups of interbreeding individuals reproductively isolated from other such groups. Evolution involves both adaptation and species formation. These are not the same process. Adaptation can account for change through time but does not inevitably lead to proliferation of species.

3.5 FURTHER READING

Futuyma, D.J. (1986) *Evolutionary Biology.*
Mayr. E. (1970) *Populations, Species and Evolution.*
Ridley, M. (1993) *Evolution.*

PART TWO
Population processes

Populations 4

4.1 GROWTH, REGULATION AND INTERACTION

Organisms come into being and in due course, die. Between these two events they have a chance to grow, move, interact with other individuals and reproduce. Sometimes individuals are fairly evenly distributed in space; in other species or circumstances they may vary markedly in density, so that there are local patches of great abundance interspersed by areas where individuals are rare. In either case there may be quite elaborate behaviour patterns which generate territories and regulate the movement and comportment of the individuals. One has only to think of some familiar plants and animals – oak trees, primroses, domestic cats, wood mice, herring gulls, *Paramecium* – to realize that population structure may be immensely complex and varied and that the proper understanding of any one of them could take a lifetime. As so often happens, we are forced to make radical simplifications in order to create an intellectual structure within which questions may be posed and generalizations made and tested.

4.2 RATES AND GENERATIONS IN SINGLE POPULATIONS

In order to see the wood for the trees, the average properties of individuals in populations are often described as rates at which things happen. Instead of following individual life histories we may refer to birth rate, death rate, rate of diffusion, etc. In an organism with continuous birth and death rates, of which the most familiar is man, these rates would be expressed per unit time. Many plants and animals have non-overlapping generations, however, and it is practical to describe birth, death and movement as the net amount taking place within a generation, which, having been completed, then gives rise to a new generation. This pattern of non-overlapping generations is often assumed when developing theoretical models. Yet another useful simplification is to assume that the distribution consists of well-delimited patches of high density with low levels of contact between them. Then the patches of high density can be considered to be populations, for each of which we may measure a birth rate, death rate, etc., while the interconnection may be measured as migration rate, the fraction of one population moving to the next. These populations are sometimes called demes.

In considering the theory of gene frequency change, a model of this kind will be used. A species will be assumed to consist of individuals with non-overlapping generations distributed as well-defined populations (demes) each with its own gene frequency (an average property of that population) and migration between them (Figure 4.1).

Many species actually do live like this; it is often true of annual insects with specific food plants or other ecological requirements. Population geneticists use the model in order to get started with a theoretical system which will allow predictions to be made. They do not believe it to be universally applicable and modify it where particular considerations have to be taken into account.

It helps to clarify the different factors involved in population growth and regulation if they are described in algebraic terms. Since algebraic descriptions will be used frequently in the ensuing chapters, it is worth justifying them now. Mathematical language provides a way of writing precise statements about processes and events. Once the problem has been formally set up it may be possible to deduce unexpected

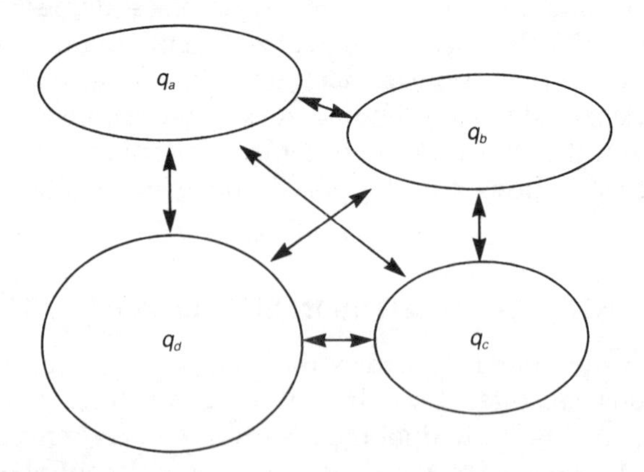

Fig. 4.1 A formal system for studying evolution in populations. Each population is a panmictic unit, within which every individual has an equal chance of mating with any other of the opposite sex. The arrows indicate that migration may take place between any pair of units, although the rate is likely to be dependent on distance. Generations may be non-overlapping, in which case each generation of adults gives rise to a certain number of offspring and then dies. Time is measured in generations. If generations are overlapping then it is necessary to record age-specific birth and death rates per unit time. A generation is then the time required to reach the age at which the mean number of offspring is produced.

consequences. More importantly, perhaps, the mathematical statement removes the ambiguity which is often present in everyday language. In formulating the problem, we are forced to decide which assumptions are necessary and whether or not they are sufficient to describe what is intended. Mathematicians sometimes chide biologists for meddling in a form of literacy (the algebraic one) at which they may not be adept. It is nevertheless worth trying; the worst outcome is to be manifestly wrong, the alternative risk is to be inpenetrably obscure. In a later section we will discuss a proposition which says that the occurrence of deleterious mutations reduces the average fitness of individuals in a population. Is this necessarily true? The answer depends on how fitness is achieved, which is not immediately obvious but is clarified by mathematical formulation. If average fitness is defined in a certain way it may also be shown that no matter how deleterious a given mutant is, the average effect is always the same. This is an example of an insight which would be very difficult to arrive at without algebra, but which follows rather simply when it is used.

To return to population growth, change in numbers can be described by writing,

$$N_{n+1} = N_n R$$

R is the net rate of increase; in a bisexual population with non-overlapping generations it is the average number of offspring produced per pair. The subscript n represents the generation to which the population refers. If we start from generation 0, this recurrence relation allows us to write the equation,

$$N_n = N_0 R^n$$

which describes change in numbers in a population with non-overlapping generations exhibiting unlimited increase. This may be modified for any time interval t by writing

$$N_t = N_0 \mathrm{e}^{rt}$$

The value $r = \ln R$ is referred to as the intrinsic rate of increase, and the equation would apply to a population with overlapping generations if there was unlimited increase. Given this equation, it can be shown that the rate of change in numbers with time is,

$$\frac{\mathrm{d}N}{\mathrm{d}t} = rN_0\,\mathrm{e}^{rt} = N_t r$$

The change goes on increasing as N increases. Change in a real population must come to a stop eventually. It has to be represented by some form of equation in which $\mathrm{d}N/\mathrm{d}t$ has its maximum value when N is very small and becomes zero when N has reached some upper limiting value. The form which has become embedded in the ecological literature is the logistic equation

$$\frac{dN}{dt} = Nr\left(1 - \frac{N}{K}\right)$$

where the N values refer to time t. This was first used in the nineteenth century and has continued to be employed as a kind of archetypal description of sigmoid growth ever since. The part in parentheses ensures that the rate of increase declines linearly as N increases from a negligible value until dN/dt becomes zero at a saturation value of $N = K$. The curve is sigmoid and symmetrical about $N = K/2$. Numbers can usually be related to a space occupied, so that K can be considered to be a value indicating a density of individuals per unit area of land or volume of water, or perhaps of forest canopy. The restraint exerted is said to be density dependent.

The logistic equation has been very influential in allowing people to visualize the way in which population growth rate reacts to density. It may, nevertheless, be misleading in two respects. In the first place, the notation suggests that r, the intrinsic rate of increase, and K, the saturation density, are independent of each other. This may possibly be so, but we can easily imagine that a greater value of r may sometimes allow a greater saturation density to be attained. The equation can be written in a more general form as

$$\frac{dN}{dt} = Nr - bN^c$$

The logistic is the specific case where $c = 2$ and $b = r/K$. Secondly, being a differential equation, the logistic assumes an instantaneous response on the part of the population to a change in numbers. Again, an effectively immediate response may occur, but it is much more likely that there will be time lag between cause (population increase or decrease) and effect (adjustment of growth rate). With the logistic, a population starting at a value of N below K can never exceed K, and whether we choose to start below or above the equilibrium level, the population will approach it asymptotically. In practice, populations may overshoot, and oscillate about, their apparent equilibrium densities, and this is especially apparent when generations do not overlap.

Both limitations may be removed by writing the recurrence relation

$$N_{n+1} = N_n R\left(1 - \frac{N_n}{K}\right)$$

This is different from, although conceptually related to, the logistic equation. It is often useful to rewrite the relation as

$$\begin{aligned}\Delta N &= N_{n+1} - N_n \\ &= \frac{N}{K}(RK - K - RN)\end{aligned}$$

An equation of this type is called a difference equation. One advantage of the form is that it is a direct measure of the change taking place. Since N_{n+1} is expressed in terms of N_n, all the terms refer to the same generation. Equilibrium occurs when $\Delta N = 0$, at which point

$$N = \frac{K(R-1)}{R}$$

The change in ΔN with change in N depends on the rate of increase. At the equilibrium value of N,

$$\frac{\mathrm{d}\Delta N}{\mathrm{d}N} = 1 - R$$

From this equation it can be shown that, for values of R between 1 and 2, there is monotonic approach to the equilibrium value. When R lies between 2 and 3 there are damped oscillations, and above 3 the oscillations increase and become more irregular with progressive increase in R until chaotic distributions are reached. Fluctuations in numbers therefore need not always be caused by extrinsic factors (weather, enemies and so on), but may themselves be a consequence of a high capacity for increase in a limited but constant environment. The likelihood of fluctuation may also depend on how the density-dependent effect works. The restraining effect is described as a contest when the resource depletes steadily as numbers increase. Sometimes a sudden crisis point occurs, however. If enough food is available for all the eggs laid by a blowfly on a decaying carcass, then no competition need occur. If too many eggs have been laid then nearly all the larvae may die when half grown. The effect is catastrophic, and the competitive pattern is a scramble for resources. Fluctuations in numbers are likely to be wilder when scrambles can occur.

The equation used here is still a simplified and abstract representation of any real population, but considerably nearer to reality in its behaviour under different values of R and N. One step further would be achieved by including a term representing mortality independent of density, which fluctuates at random about some mean value. It may be reasonable to represent the change by

$$N_{n+1} = N_n R\left(1 - \frac{N_n}{K}\right) - mN_n$$

where m is a random normal deviate with variance V_m.

Very often, organisms have overlapping generations and distinct life stages. We are accustomed to extreme generation overlap in human populations. Insects and marine crustacea have obvious larval and adult forms subject to different kinds of environmental constraints. An insect such as *Drosophila*, with a short life cycle in an annually fluctuating environment, may start the year with discrete life stages, which then show progressively greater overlap as the year proceeds, the weather improves and the life cycle speeds up. Many land snails take 1 or 2 years to become mature, then live for several more years. These patterns result in different growth processes. To investigate the way in which they operate it is convenient to divide the life span into a series of stages and to define a rate of survival and a rate of reproduction for each stage. We may have a population divided into four age classes, x_1, x_2, x_3 and x_4. Each class has a fertility of f_1, f_2, f_3 and f_4, and each individual has a probability of survival to the next age class of p_1, p_2 and p_3 (if they only live through four of these age classes, then $p_4 = 0$). To get the numbers in each age class one unit of time later, we multiply the probability of survival by the number available to survive, i.e.

$$x'_2 = x_1 p_1$$
$$x'_3 = x_2 p_2$$
etc.

and to get the number in the youngest age class we must multiply the number in each class by its fertility, so that

$$x'_1 = x_1 f_1 + x_2 f_2 + x_3 f_3 + x_4 f_4$$

A more economical way of expressing this is in the form of matrices. They were first applied to population problems by P.H. Leslie, so that the transition matrix representing fecundity and mortality is called a Leslie matrix. Population changes are represented by multiplying this by a vector showing numbers in each age group, as follows,

$$\begin{bmatrix} f_1 & f_2 & f_3 & f_4 \\ p_1 & 0 & 0 & 0 \\ 0 & p_2 & 0 & 0 \\ 0 & 0 & p_3 & 0 \end{bmatrix} \begin{bmatrix} x_1 \\ x_2 \\ x_3 \\ x_4 \end{bmatrix} = \begin{bmatrix} x'_1 \\ x'_2 \\ x'_3 \\ x'_4 \end{bmatrix}$$

The upper row of the square Leslie matrix represents fecundity at successive stages, and is analogous to the R in the equation for non-overlapping generations. The subdiagonal, p, row represents the probability of surviving from one stage to the next. The x column vector represents numbers of individuals in each age category at a given time, and multiplication of the matrix by the vector provides the numbers in the next stage, the x' column. If this were to be applied to a long-lived snail, such as *Cepaea nemoralis*, 4 years would be indicated, and f_1 and f_2 would be zero, since the animals are juvenile and do not reproduce until the third year. As adults, they produce perhaps 50 eggs per individual (f_3 and f_4 equal to 50). Survival from egg to first year (p_1) may be only about 0.01, while survival from one year to the next in older animals is about 0.5. Using these values we could start with any values of x_i. So long as adults were included numbers would then increase from year 0 to year 1, the x' values are then substituted for the x values and the process repeated. Oscillations in numbers would be set up until the population settled to a steady age distribution, whereupon the ratio of totals in successive column vectors indicates the rate of exponential increase (analagous to R in the simple equation). To make such a population approach an equilibrium, density-sensitive functions have to be introduced to reduce f values and/or p values as the total increases. If the population is made in this way to attain an equilibrium, then disturbance from it may readily be made to result in stable limit cycles fluctuating continuously between a maximum and a minimum density. Quite complex patterns of change may therefore be generated even with such a comparatively simple life cycle as that of a land snail. In human populations, short-term fluctuations due to changes in birth and death rates make it very difficult to predict future needs in relation to energy consumption, housing, school places, etc.

This digression into patterns of population growth and regulation would not be complete without a brief consideration of the way populations of different species may interact. The interaction may be between species at the same trophic level using similar resources, or it may be between species at different trophic levels, when one exploits another, as predators do prey or parasites, hosts.

4.3 SPECIES INTERACTIONS

Equations like the logistic indicate that additional individuals in a population have an inhibiting effect on further increase in numbers, and that this inhibition becomes more powerful as total population size increases. When more than one species lives in the same area the inhibition may be exerted by individuals of the same or of a different species. To describe the processes involved it is necessary to have equations representing each species, which interact with each other. Thus, two species could be represented as

$$N_{1,n+1} = N_{1,n}R_1\left(1 - \frac{N_{1,n} - a_2 N_{2,n}}{K_1}\right)$$

$$N_{2,n+1} = N_{2,n}R_2\left(1 - \frac{N_{2,n} - a_1 N_{1,n}}{K_2}\right)$$

Change in numbers in each is affected by its own and the other species. Equations of this type are associated with the name of G.F. Gause, the Russian biologist who investigated species competition and the concept of the niche in the 1930s. The species compete for resources, and the values of R, K and a determine the type and amount of competition. Given such a system, there are four possible outcomes. Either species 1 or species 2 wins while the other loses, or an equilibrium is set up. The equilibrium may be stable, whereupon both species can coexist, or unstable, in which case the species with the highest starting number wins and the other becomes extinct. If $a_1 = a_2$ then the only possible outcomes are the first two.

If a_1 and a_2 both equal unity, then the species with the highest value of K will be the winner. The competition concerns the way in which a limited resource is utilized, and is said to be exploitative. If the K values are the same, the outcome depends on the a values. Some mechanism is implied, which results in interference with one species by the other. Putting both processes together and using differential equations, the conditions for the four types of outcome are:

1. $a_1 > \frac{K_2}{K_1}, a_2 < \frac{K_1}{K_2}$ species 1 wins

2. $a_1 < \frac{K_2}{K_1}, a_2 > \frac{K_1}{K_2}$ species 2 wins

3. $a_1 < \frac{K_2}{K_1}, a_2 < \frac{K_1}{K_2}$ coexistence

4. $a_1 > \frac{K_2}{K_1}, a_2 > \frac{K_1}{K_2}$ either survives depending on initial numbers

Using the recurrence relations displayed above, we have to multiply K_1 by $(R_1 - 1)/R_1$, and K_2 by $(R_2 - 1)/R_2$. The constraints are relaxed if we add terms representing random independent mortality, since these mortality coefficients may counteract or reinforce the a and K effects. If more than two species coexist and interact then the equations may be extended, with more a coefficients and different R and K values for each species. For k species,

$$N_{i,n+1} = N_{i,n}R_i - N_{i,n}\frac{R_i}{K_i}\sum_{j=1}^{k} a_{i,j}N_{j,n}$$

where $a_{i,j} = 1$ when $i = j$.

Species may avoid competitive interaction if they can adapt by undergoing character displacement, so as to exploit different resources. When species come into direct competition, rarification as a result of random mortality increases the chance of coexistence, and so do behavioural patterns such as patchy or contagious distribution over the resources. Groups of similar, closely related species which coexist and probably compete, are called guilds. An example is found in the genus *Drosophila*, where several species live together and feed on ephemeral resources which are usually fermenting fruits, sap fluxes, decaying plant material, fungi or flowers. Estimates of guild sizes in various regions are given in Figure 4.2. Observations on field data and simulation studies suggest that spatial patchiness is important in permitting coexistence in these insects.

If an animal feeds on a plant species or one animal feeds upon another then the interaction may also be described by a pair of equations. The first such descriptions were the Lotka–Volterra equations, developed in the 1920s and applied to the interaction between parasitoid insects and their insect hosts. There was much interest in the possibility that interaction between species at different trophic levels was the key to understanding cycling of animal numbers. If so, then it may also explain the control of numbers, so that generally species oscillate about equilibrium levels, set not by density-dependent response to the physical environment but by density-sensitive species interactions. This view was advocated by the Australian biologist A.J. Nicholson. It evoked the reaction that perhaps most species were not controlled at all, in the sense of being subject to precise feedback reactions, but fluctuate within limits set by their capacity to increase, on the one hand, and accidental, weather-induced mortality on the other. As before, theoretical investigation of the proposed systems showed what kinds of outcomes were possible and what assumptions were necessary to make them at all realistic.

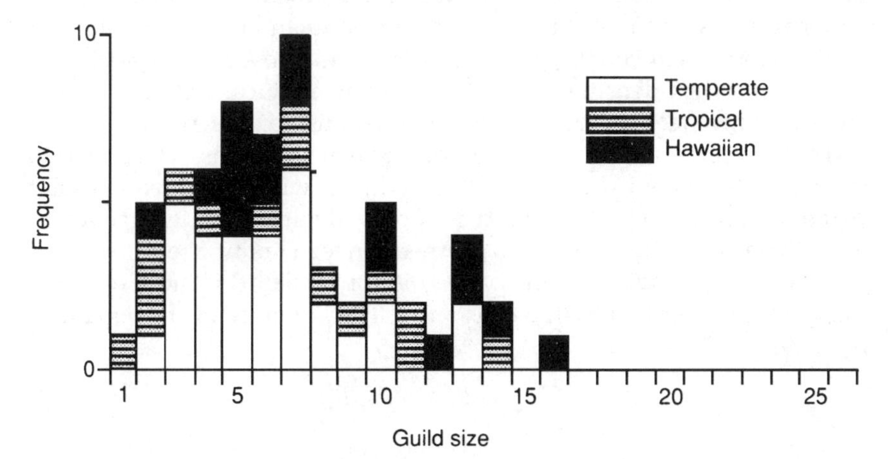

Fig. 4.2 Frequency histogram of observed guild sizes in groups of drosophilids from different regions of the world. Hawaii is a centre of adaptive radiation for the group. There is no apparent difference in pattern of guild size between localities in temperate or tropical regions or from the island archipelago. Using empirical data on distribution of competition coefficients and on degree of aggregation, the authors were able to simulate coexisting groups of species with similar numbers and distributions to those observed. Reproduced with the permission of Blackwell Science Ltd from *Organization of Communities Past and Present* (eds J.H.R. Gee and P.S. Giller), Shorrocks and Rosewell (1987).

Using the Lotka–Volterra equations, the rates of change in host and parasitoid populations may be written as:

$$\frac{dH}{dt} = Hr_H(1 - B_H P)$$

$$\frac{dP}{dt} = -Pr_P(1 - B_P H)$$

where r and B refer to the rate of increase and equilibrium level of host and parasitoid, respectively. At equilibrium $1 - B_H P = 0$ and $1 - B_P H = 0$. In theory, these equations produce continuous cycles of numbers in the two species. In practice, however, they are not stable; accidental perturbations cause the cycles to increase in amplitude, so that the species would become extinct. The history of modelling predator–prey and parasite–host interactions has been one of finding out what modifications are necessary to introduce a realistic degree of stability. A development for the Lotka–Volterra model is the one used by G.C. Varley and G.R. Gradwell, who represented the change from one generation to the next as

$$H_{n+1} = H_n^b Re^{-aP_n}$$

$$P_{n+1} = H_n^b (1 - e^{-aP_n})$$

As before, R is the rate of increase of the host. By adjusting the values of the constants a and b these equations can be made to yield a range of patterns similar to those obtained for single populations with time-lagged density-dependent control, ranging from damped oscillations to chaotic behaviour. When densities are reduced below a critical level the system shows positive feedback, so that the populations become extinct.

This system is still not a good model for real situations, which tend to show greater stability than can be produced with the three constant parameters R, a and b. The stability is improved if it is assumed that R varies with density (due, perhaps, to competition for a limiting resource), or if the searching behaviour of the parasitoid (or predator) is more sensitive than can be described by the constant a. The structure of the equations becomes

$$H_{n+1} = H_n R f(H_n) f(H_n, P_n)$$

$$P_{n+1} = H_n\{1 - f(H_n, P_n)\}$$

where $f(H_n)$ is a function describing density-sensitive reproduction in the host or prey population and $f(H_n P_n)$ describes the density-sensitive interaction. Many types of behaviour may modify the latter function. One of these is non-random distribution of the prey species. If the prey are contagiously distributed, the system is more stable than if they are uniform or random. It may be that we can conclude that predator–prey interactions favour clumped distributions, as does competition at the same trophic level; certainly these distributions are very common in nature. Problems of biological control have been modelled successfully using this approach.

4.4 THE PANMICTIC POPULATION

Besides locating its boundaries, another kind of criterion may be employed in defining a population. An individual born at a given spot has a greater chance of mating with a member of the other sex which originated close to it than one which originated far away. In a continuous distribution it is therefore possible to conceive of an area within which every individual has an equal chance of mating with any other of the appropriate sex, whereas beyond this area the probability of mating declines. The area within which there is random probability of mating is called a panmictic unit and the condition of random mating is called panmixia. Again, some species are distributed discontinuously such that each population is a panmictic unit, while in others the concept can be no more than an abstraction. It will be found to be a useful one, however, when investigating patterns of geographical distribution.

4.5 NATURAL SELECTION

Evolution proceeds by adjustments to the properties of individuals living in populations. We have next to consider how these adjustments may occur. The most substantial force causing modification of lineages of individuals is natural selection. Not all changes through time need be selected, but selection brings about increase in adaptation and is in this respect the guiding agency in evolution. According to the Darwinian model, natural selection results from competitive interaction driven by the innate capacity to increase. Some of the variation on which selection acts is inherited, so that selection leads to evolution. Notice that natural selection does not necessarily result in evolution, any more than genetic change necessarily implies selection; it will only do so to the extent that the traits concerned are inherited. What it does do is to put evolution firmly into an ecological context. On this argument, evolution appears to be the result of density-regulated competitive interactions. We can see that this does not always have to be the case for the following reasons. First, some deleterious conditions, such as the severe defects spina bifida, anencephaly or phenylketonuria in man, are likely to be selected against whatever the density or prevailing ecological conditions. Darwin was looking for the conditions which favour advantageous forms, rather than those which are clearly deleterious, but it is useful to note that relative frequency change is not inevitably density regulated. Second, during a phase of exponential increase in numbers there will be no competitive interactions, but one form could reproduce more rapidly than another and so increase in relative frequency. Evolution may therefore take place without competition. Nevertheless, much change in relative frequency is due to differential survival. We shall therefore next look at some patterns of natural selection through differential survival before getting on to the problem of inheritance and evolution.

4.6 PHENOTYPIC VARIATION

Organisms in a population vary in almost all their measurable characteristics – in size, shape, colour, longevity, concentration of enzymes, ability to withstand cold, etc. Any feature so measured is part of the phenotype of the organism, and the phenotype is generated by an array of inherited and internal and external environmental effects. Very often these effects are individually small but numerous and independent in their action. If so, then the variation between individuals is normally distributed. Sometimes several factors acting in the same direction, for example, to increase the size of a structure, have a multiplicative instead of an additive effect. In that case the distribution will show an extended tail in that direction, i.e. it will be skewed. Transformation of the scale in which the variable is measured from arithmetic to logarithmic will then make the distribution more symmetrical (Figure 4.3).

When information on variable characters is collected, a standardized method of summarizing it is usually employed. This involves specifying the central tendency of the distribution of the data, and its spread. The central tendency is most often expressed as the mean value and the spread as the variance, which is the average value of the squared deviations from the mean. For a set of n values x_i the mean is

$$\bar{x} = \frac{\Sigma x_i}{n}$$

If $\bar{x}$ is calculated from the data, then the variance, usually symbolized as V or s^2 is

$$V = s^2 = \frac{\Sigma(x_i - \bar{x})^2}{n-1}$$

The denominator is $n - 1$ because at least two values are required to measure variation. If the mean $\bar{x}$ is fixed, for instance when movement is measured from a fixed starting point, then

$$V = \frac{\Sigma(x_i - \bar{x})^2}{n}$$

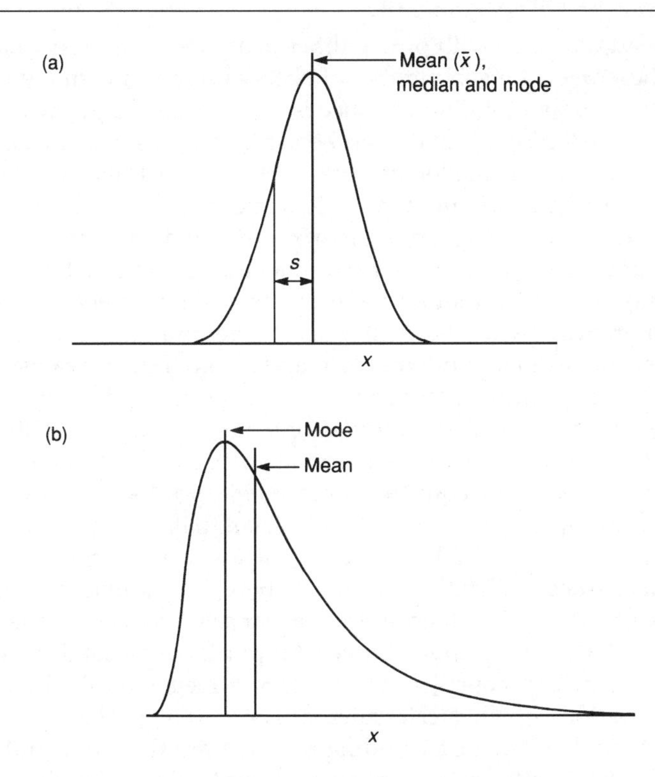

Fig. 4.3 Natural phenotypic variation. Samples from a population varying in some trait. (a) Variation due to random effects of many independent variables with small additive effects. These tend to conform to normal distributions. The standard deviation *s* is shown. (b) Variation due to effects of many independent variables with small multiplicative effects. These generate skewed distributions, which may be made more like normal distributions by transforming the scale to log *x*.

The number of occurrences of each value x_i may sometimes be measured as a frequency f_i. The sum of f_i is 1, and the variance becomes

$$V = \Sigma f_i(x_i - \bar{x})^2$$

The standard deviation, *s*, is the square root of the variance.

The variance is a parameter of the normal distribution, along with the mean. That is to say, a normal distribution may be generated if we know the mean and the variance. The standard deviation then measures a distance from the mean which includes 68.25% of observations. This fact is used in statistical tests of significance. The variance may be used as a legitimate measure of variation, however, even if the data are not normally distributed.

Another distribution, which is distinct from the normal distribution, even though it sometimes resembles it in shape, is the binomial distribution This occurs when data can only exist in one of two states (male/

female, head/tail, etc.) with probabilities p and $1 - p$. For a set of n observations the expected mean number of observations of the first type is pn, and it can be shown that the variance is $p(1 - p)n$. For a given value of p the variance gets proportionally larger as the sample size increases. For a given sample size it is smaller at values of p near 0 or 1 than at central values, reaching its maximum at $p = 0.5$. Sometimes p is very small. This would be the case, for example, if it represented the probability of a mutation at a locus, where $1 - p$ represents the probability that the locus has not mutated. If we plotted the frequency of mutants at a sample of loci in different individuals, most would have no mutated genes, a few one, a very much smaller number two and so on. If the mutants occur at random the distribution generated is called a Poisson distribution. Like the binomial, only two states occur (mutated or not mutated in the example). When p is very small, a binomial distribution would have its variance $p(1 - p)n$ nearly equal to its mean pn, since $1 - p$ nearly equals 1. The variance of a Poisson distribution is equal to its mean.

In comparing different kinds of organisms, a scaling consideration must be introduced. Variation in ear size between individual mice must obviously be smaller than the variation in ear size between elephants, just because the latter are so much bigger. Proportional variability need not differ, however. This consideration leads to a measure called the coefficient of variation, which is calculated as $CV = 100s/\bar{x}$. The same kind of adjustment could be made by finding the standard deviation of the logarithms of x. The coefficient of variation expresses the variability as a percentage of the mean value. It turns out that for a wide range of characters of animals of different sizes the coefficient of variation tends to lie between 5% and 10%. Organs such as antlers in deer, which have a social as well as a mechanical function, produce higher values of around 15–20%, as do repeated characters such as the ray florets of Michaelmas daisies. Birds, which are subject to strict mechanical constraints because of the need to fly, are likely to exhibit smaller coefficients of variation than terrestrial animals.

A common pattern of selection within any one generation is for a trait to start with a normal distribution and a coefficient of variation of some 5% and then to suffer selective elimination before reproduction occurs.

4.7 SELECTIVE ELIMINATION

If the variable considered is unaffected by change in size, such as, for example, the number of scales in a row on the ventral surface of a lizard, then the distribution in a sample at birth may be compared directly with the distribution in the sample just before reproduction. If the variable is one, such as leg length, which would increase as size increases, then some sort of scaling is required. Usually, the coefficient of variation or the logarithm of the variable would be appropriate. In either case, the process starts with one distribution and ends with another, which is smaller and may have a different shape.

Some of the mortality will be independent of the variable measured, and for that reason, non-selective. The first step in measuring selection is to remove this non-selective fraction. Diagrammatically, this may be done by multiplying up the values of the lower curve by the same factor for all values of x until the lower curve just touches the upper one (Figure 4.4). The value of x at which the two curves touch is the optimum value x_0, at which the minimum mortality is experienced. The area between the two

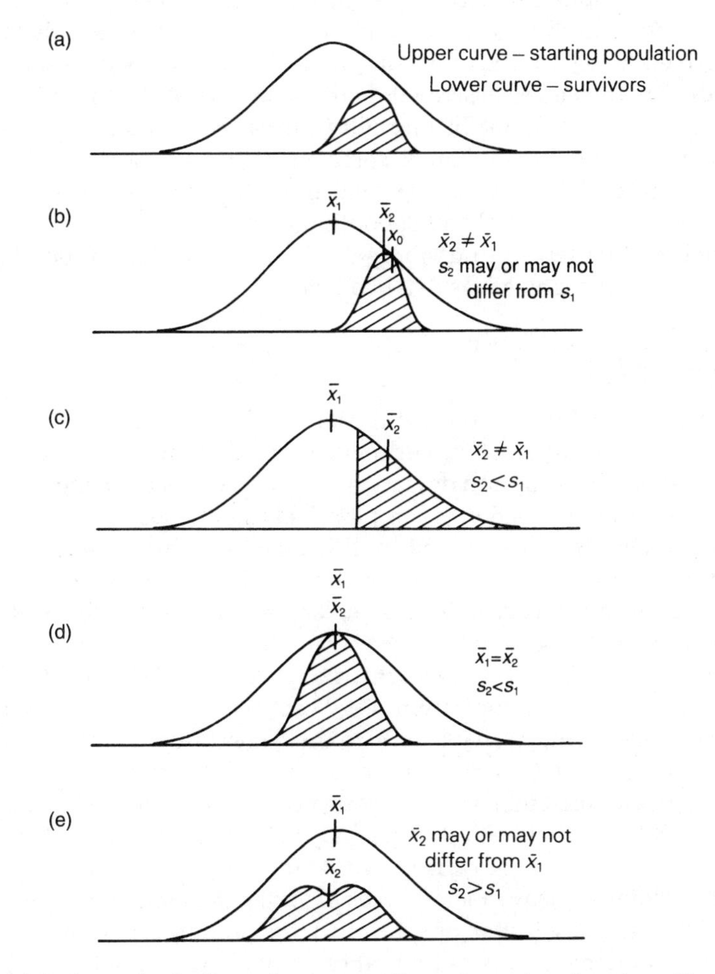

Fig. 4.4 Patterns of selection acting on a continuously varying phenotype. Shaded sections indicate surviving fraction of the populations: (a) and (b) directional selection; (c) threshold selection; (d) stabilizing selection; (e) disruptive selection.

curves then represents the mortality suffered by individuals with other values of x as a result of their deviation from x_0.

It is now possible to formulate a measure of the total amount of selection which is being imposed. This is the intensity of selection, I, and

$$I = \frac{\text{difference between optimum and average survival}}{\text{optimum survival}}$$

$$= \frac{\text{area between curves}}{\text{starting area}}$$

This measure was first proposed by J.B.S. Haldane and much extended by L. Van Valen. Applied strictly in the form described the calculation of I would produce an overestimate. Mathematical formulae have therefore been derived, based on the initial and final means and variances. Notice that in the circumstances described there can be no selection if there is no mortality, and the amount of mortality limits the amount of selection which may operate. When there is some mortality, the intensity of selection is not related directly to the total mortality but to the difference between that of the optimal and suboptimal types.

Three broad patterns of possible selective modification of the phenotype have been recognized, as detailed below.

4.7.1 DIRECTIONAL

Figure 4.4(a) and (b) illustrates directional selection, in that the mean value for the distribution of the survivors is greater than that for the starting population. The standard deviation may or may not be changed in the process. Directional selection is comparatively rarely observed in nature. A striking example was reported by P.R. Grant and his colleagues who studied variation in the Galapagos finches belonging to the genus *Geospiza*. These birds vary in beak size, length and depth of beak, determining the size of seed which they can eat. In 1977 on one of the Galapagos islands *G. fortis* was caught in a drought which limited the food supply. As seeds of the optimum size were used up only larger ones remained, which were available only to the larger birds. Selective starvation therefore occurred, and in that season the bill length in surviving birds was greater than that of the starving population by 6%. A high correlation of bill length between parents and their offspring shows that this directional selection can be carried over to future generations.

Where mining of heavy metals takes place the deposition of toxic spoil imposes directional selection on plants. At first nothing will grow at these sites, but a number of grasses and other plants have variants which are heavy metal tolerant, and these begin to colonize the new habitats. In time an equilibrium is established, with a tolerant population on the tailing, surrounded by non-tolerant populations, with migration of pollen and seeds between them.

Figure 4.4(c) shows a pattern of directional selection often imposed in artificial selection programmes. Only individuals which achieve a value above a certain threshold are permitted to breed, so that this pattern may be termed threshold selection.

4.7.2 STABILIZING

Under stabilizing selection, the optimum value has already been achieved and selection serves to eliminate extreme variants (Figure 4.4(d)). Several famous examples occur in the biological literature. One concerns a study of birth weights of babies born in a London hospital. The average survival over the first 4 weeks of the total sample studied was 95.9%. In the middle of the birth weight distribution lay the optimal class with a birth weight of between 7.5 and 8.5 lb (3.37–3.82 kg), and these babies had a 98.3% survival rate. Using the formula for selective intensity discussed above, this indicates an intensity of (0.983 – 0.959)/0.983 or 2.4%. The selection which operates serves to eliminate extreme deviants, whether they are particularly large or particularly small. It should be noted in this case that a proper study of the system would have to take a number of factors into account. Thus, it could well be advantageous to a baby to be as large as possible at birth, while its mother has the best chance of giving birth without complications if it is small. At least two conflicting forces therefore operate to determine optimal birth weight.

At the end of the nineteenth century, two groups examined evidence for stabilizing selection. In the USA, Hermon Bumpus published an article entitled 'The elimination of the unfit as illustrated by the introduced sparrow'. In a flock of English sparrows (*Passer domesticus*) disabled by cold one winter he was able to show that survivors had a smaller variance in body proportions than the flock as a whole. Subsequent analyses have extended this work. In Europe W.F.R. Weldon collected a sample of the snail *Cochlodina laminata* from the castle of Brescia in Italy and showed that variance in whorl dimensions of the shells was greater in juveniles than in the same whorls of adults, thus indicating selective elimination. This approach has been greatly extended in modern studies by Berry and Crothers of the dogwhelk *Nucella lapillus*. A detailed study was carried out on the number of chaetae on the sternopleural plate in *Drosophila melanogaster* by Linney, Barnes and Kearsey. They collected eggs from a stock and raised them either at low densities at which 80% survived or at high densities, when only 5–8% survived. The ratio of survival at the two densities was then worked out for flies of each chaeta number. It was apparent that flies with extreme chaeta numbers were much more susceptible to the effects of increasing density than those with numbers near the mean. In newts, G. Bell measured head and body length in wild populations, and used total length (including tail) as a measure of age. The coefficients of variation decreased with increasing total length. This did not occur in laboratory controls, which indicates stabilizing selection acting in the field.

4.7.3 DISRUPTIVE

Figure 4.4(e) shows a situation where there is disruptive selection. Starting from a distribution with a single mode, selection favours two new modes. Like directional selection, this is likely to be a transient phase in the history of a given system, because one or the other mode may be expected to predominate. An exception is mimetic polymorphism in butterflies, where sets of linked genes control either one mimetic phenotype or an alternative one. Mixtures of characters of the two mimetic types would be disadvantageous and have a lower fitness than either of the two optima. Examples of mimetic polymorphisms are discussed in section 9.5. The most detailed work on disruptive selection has been carried out by J.M. Thoday and his colleagues, studying sternopleural chaeta number in *Drosophila*. They have shown that disruptive selection can lead to rapid divergence towards two modes, which, given the right conditions, will both be maintained in a population.

4.8 CALCULATION OF INTENSITY AND SELECTIVE VALUE

Most of these examples can be discussed in terms of phenotypic variation, although we know that genetic variability may be involved. Many selective situations, however, refer to two or three classes of individuals, rather than to continuous variation. Can selection then be represented in the same way? Suppose we start with two classes of individuals, A and B, in the ratio 0.6:0.4. Half of class A survives and three-quarters of class B, so that in relation to the starting number we are left with 0.3 in each case. To calculate the intensity of selection, we have an optimum survival of $3/4 = 0.75$ and an average survival of $(3/4 \times 0.4) + (1/2 \times 0.6) = 0.6$. Therefore,

$$I = \frac{0.75 - 0.6}{0.75} = 0.2$$

The intensity of selection is 20%. Another quantity which tells us something useful about selection is the fitness of class A relative to class B. If the fitness of B is set at unity this is $w = 0.5/0.75 = 0.667$. A measure of this kind is called the relative fitness or selective value of A. When there are two or a few classes of individuals the mean fitness, $\overline{w}$, is sometimes also calculated. In this case $\overline{w} = (0.6 \times 0.667) + (0.4 \times 1) = 0.8$. Selective values and relative fitnesses will be used extensively later. The maximum value $\overline{w}$ could have is unity, when all individuals were assigned a selective value of 1.0. As the example shows, when the maximum fitness is 1, then $\overline{w}$ is equal to $1 - I$.

4.9 LIFE CYCLES

The last part of the Darwinian statement of natural selection states that evolution only occurs if some part of the variability is inherited. Inheritance implies the passage from one time to another, or one generation to the next. The consequences of natural selection therefore depend on how characters are inherited.

Reproduction may be asexual, for example, by binary fission or vegetative reproduction. Almost no organisms reproduce like this to the exclusion of all other ways, but they may do so for long periods of time. Amoebae are exceptional in being known to undergo only binary fission. *Paramecium* reproduces by fission as long as conditions are satisfactory, but from time to time goes through sexual reproduction by conjugation. The English elm *Ulmus procera* is, or was before the spread of Dutch Elm disease, a gigantic asexual clone which spread through the country along hedgerows and through copses by means of the production of suckers. The lack of variation between the different stands may have increased the disastrous consequences of the disease.

When there is vegetative reproduction and a range of clones available, interclonal selection may take place. A good example, studied by A.H. Charles, can be seen in rye grass, *Lolium perenne*, seeded in swards for cattle. About 750–1600 seeds per square metre are used. About 20% fail to survive the first year, as a result of failure to germinate, attack by cattle, etc. Only about 50% get to 4 or 5 years and 5–10% manage 40 years. When a good many plants are still present, there are a few large and many small ones, so that the size distribution is strongly skewed, but as the number drops the smaller ones are lost and the skewness is reduced. The survivors tend to be prostrate, late flowering and short leaved; they are the ones which are adapted to survive the effects of grazing.

This is a very simple pattern of selection, to which the methods of measurement discussed above could be applied. If a sexual reproduction process occurs then it is necessary to consider the breeding system. The mature organism may be haploid, diploid or polyploid. The pattern usually discussed is that where the organism is diploid and bisexual (dioecious), probably because this is the pattern we have in man and common domestic animals. Sex is determined by the segregation of sex chromosomes and autosomes, so that there are sex-linked and autosomal genes. This is the pattern we will assume in the following pages unless otherwise indicated. The cycle of reproduction has the following components:

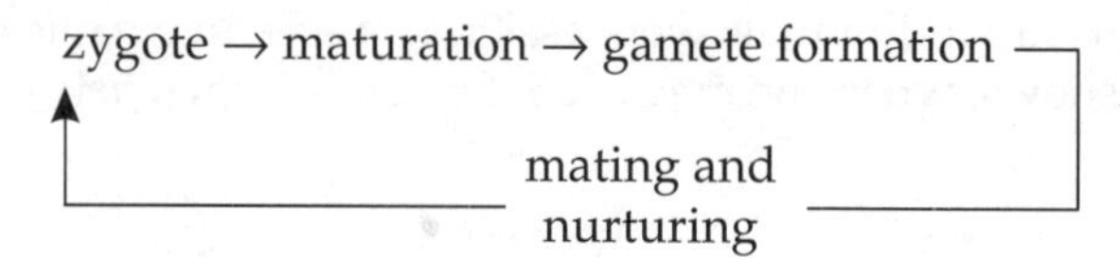

The term nurturing is intended broadly to cover any process, like placenta formation or parental care, by which individuals in one generation contribute to the support of another.

The generations may be non-overlapping, as in many annual plants and insects, in which case a change in numbers may be measured as the difference in numbers between generations (ΔN). It may be overlapping, as in man and the larger mammals, in which case the change is measured as dN/dt.

There are almost always more than two gametes formed per adult in a population, sometimes immensely more. Adjustment is therefore needed to achieve a stable population. This may be brought about in the following ways:

1. Stage: zygote/adult – Adjustment: mortality;
2. Stage: gamete production – Adjustment: (i) Production, (ii) Mortality;
3. Stage: mating – Adjustment: non-random;
4. Stage: nurturing – Adjustment: (i) Ineffective, (ii) Mortality.

All the adjustments may vary from one individual or one gamete to another, to result in selection. The gametes which pass through the process carry haploid subsets of hereditary information derived from the parents and combine to produce new mixes to form the next generation. How this process is brought about is the raw material of genetics.

4.10 SUMMARY

Natural selection takes place within populations. A picture of the dynamics of populations is therefore helpful to an understanding of evolution. Population dynamics is the outcome of a potential for geometric increase curbed by the limitation of finite resources. If there is a substantial time lag between the onset of a limiting effect and the response by the population the result may be to set up cycles or fluctuations in numbers. If individuals have overlapping generations and pass through prereproductive and reproductive stages then fluctuations become more likely, and extrinsic environmental effects increase their likelihood. Interactions between predators and prey and parasites and hosts tend to generate cycles in numbers. Species at the same trophic level may compete with each other for the same resource (exploitative competition) or each may have a negative influence on some specific component of the capacity for increase of the other (interference). These effects on numbers all tend to lead to extinction; nevertheless populations of most species are very resilient. One reason is that many of the factors influencing them act in a density-dependent manner. Another is that irregular spatial distribution of individuals in prey, predator, parasite and disease populations has a stabilizing effect.

Evolution occurs when the genetic constitution of a population is modified with time. The modification may occur at any stage of the life cycle and affect survival or capacity for increase. Populations are defined as panmictic when they are of such a size that all individuals have a chance of combining their genes with those of any other during the mating process. In such a population, selection may result from competition between individuals for resources (natural selection as discussed in the simple Darwinian theory). Disadvantageous characters will also tend to be eliminated without there being overt competition, and particularly favourable ones will tend to increase in frequency. For aspects of the phenotype which show continuous variation these selective processes may sometimes be described and measured in terms of changes in means, variances and shapes of phenotype frequency distributions.

4.11 FURTHER READING

Begon, M., Mortimer, M. and Thompson, D.J. (1996) *Population Ecology.*
Berryman, A. (1999) *Principles of Population Dynamics and their Application.*
Hassell, M.P. (1978) *The Dynamics of Arthropod Predator–Prey Systems.*
Ricklefs, R.E. (1997) *The Economy of Nature: a Textbook of Basic Ecology.*
Roughgarden, I. (1979) *Theory of Population Genetics and Evolutionary Ecology: an Introduction.*
Royama, T. (1992) *Analytical Population Dynamics.*
Williamson, M.H. (1972) *The Analysis of Biological Populations.*

Genes 5

5.1 INTRODUCTION

At this stage it is useful to consider how genes are to be treated in the context of population genetics. The gene is known from studies at three levels:

1. from how it appears in pedigrees or in breeding experiments (Mendelian genetics);
2. from how it is composed chemically and the molecules it produces (molecular genetics);
3. from how it functions in the organism to give rise to the phenotype (gene action and developmental genetics).

Like other branches of science, genetics has contributed many new terms to the language. Sometimes it seems that the subject is jargon-ridden due to a capricious inventiveness on the part of geneticists. In fact, each new term is coined to meet a need in explaining some new or baffling piece of evidence, and after a useful life of a number of years many of these terms become redundant and should be allowed to die. Many of them continue to play a valuable role, however, acting as shorthand indicators of the problems to which they relate. This short review will be hung on a series of terms dating from the beginnings of genetics to the present day.

5.2 TRANSMISSION OF INHERITED MATERIAL BETWEEN GENERATIONS

Genes are situated on chromosomes. Organisms have thousands or tens of thousands of genes, but one to hundreds of chromosomes. Genes situated on the same chromosome are said to be linked, while those on different chromosomes are unlinked. The arrangement of chromosomes possessed by an organism, which can be seen microscopically at some stages of the cell cycle, is called the karyotype.

Most animals are diploid, that is to say, each cell possesses pairs of homologous chromosomes. The fully formed organism develops from the fertilized egg, or zygote, by cell division accompanied by chromosome replication known as mitosis. Each chromosome is copied exactly to the daughter cells. Meiosis, in the course of gamete formation, ensures that exactly half the chromosome complement containing one of the homologous sets of genes (the haploid set) passes to egg and sperm, so that diploidy is maintained at reproduction.

The assortment of chromosomes in homologous sets is accomplished by physical processes initiated by the centromeres, parts of chromosomes concerned with replication. These may lie centrally on a chromosome (metacentric), at one end (telocentric) or near an end (acrocentric). During meiosis, homologous chromosomes may exchange parts by the physical process of chiasma formation, or crossing-over, so that linked genes appear in new combinations. Occasionally, breakage and misrepair at interphase causes non-homologous sections of chromosomes to become attached (translocation). Such translocations are heterozygous when first formed, and heterozygotes often experience difficulties at meiosis which result in genic imbalance leading to inviable gametes. Occasional misrepair in a single chromosome results in a deletion or an inverted segment. These may likewise result in loss of viability.

The process outlined consists of alternation of a diploid organism and a brief haploid phase consisting of gametes. The arithmetic of gene transmission requires that some such alternation takes place, but plants show more variation in the way it is brought about. The stage which bears gametes is known as the gametophyte; it is the dominant stage in algae, fungi and bryophytes (mosses and liverworts). In ferns the stage we usually notice is the sporophyte, bearing spores which develop into a reduced gametophytic stage consisting of tiny free-living plants. In seed-bearing plants this stage is reduced to microscopic structures. The female gametophyte is retained as part of the flower structure and male gametophyte is dispersed as pollen.

Thus, part of the genetical process is the distribution of genes from one generation to the next in packets which ensure constant numbers of alleles at each genetical locus and balance of activity between loci. Because of the physical nature of these packets, the chromosomes, there are restrictions on the way genes are transmitted, which sometimes play an important part in evolution.

5.3 MENDELIAN GENETICS

From study of the outcome of experimental crosses and pedigrees, it is useful to think of many phenotypic characters as controlled by segregating genes. Since they segregate we can devise combinatorial rules which govern their behaviour and express the relations of the alleles and loci to each other by a set of descriptive terms. Thus, if two unlinked loci each have a pair of alleles, one showing dominance to the other, then a cross of two double heterozygotes (called an F_2 cross) will give rise to a 9:3:3:1 ratio of four phenotypes among the offspring. Conversely, if this ratio is achieved we can, with suitable reservations, infer the conditions stated. Let us now consider what these conditions imply.

Two unlinked loci will segregate independently. If they are on the same chromosome, they are linked cytologically. If they are close together then their contiguity will ensure that they tend to be inherited together and the map distance between them is the same as the probability of recombination, or cross-over value. However, this relationship only exists for relatively low recombination frequencies or short chromosomes; if two loci are more than 50 map units apart then the probability of recombination is 0.5 and they behave in the same way as pairs of loci on separate chromosomes. Many organisms have chromosomes up to hundreds of map units long.

One allele at a locus is dominant to another (which is therefore recessive) if the phenotypic expression of the homozygote and heterozygote are identical. Thus, for example, the condition unbanded shell in the snail *Cepaea nemoralis* is dominant to five-banded. It must be emphasized that this description refers to the trait as we perceive it. Sometimes, recessiveness arises because an allele fails to function at all and the functioning allele in the heterozygote can produce sufficient gene product to compensate for this lack. Albinism in mammals or white eye (*w*) in *Drosophila* come into this category. They result from lesions in the metabolic pathways which produce the normal colour. But the situation is not usually as simple as this. Another gene in *Drosophila*, curly (*Cy*), is lethal when homozygous but in the heterozygote causes the wings to curve upwards to a greater or lesser extent. It is recessive if we study survival but dominant if we examine wings. The brown rat, *Rattus norvegicus*, is sometimes killed by feeding it bait containing the poison warfarin. In more than one natural population, this has led to the development of warfarin resistance controlled by a single gene (Rw^2). Heterozygotes as well as homozygotes for this gene survive doses of warfarin, so that resistance is dominant, but homozygotes also suffer from vitamin K deficiency. This metabolic effect of the gene is recessive.

Genes for resistance to insecticides are known in many insect species. Their effect is usually studied in experimental populations by measuring the mortality of samples of insects subject to different doses of the insecticide. In this way the resistance genes have often been shown to be codominant, in that the heterozygote is distinguishable from both homozygotes (Figure 5.1). The investigator can discriminate between the genotypes by virtue of being able to alter the environment experienced by the insects. In the wild, however, only one mean dose of the insecticide will be experienced by a population, and the gene will be expressed as a recessive, codominant or dominant depending on what that dose is. Dominance is not only a description of the phenotype rather than an intrinsic property, dependent on how the observer perceives it, but it also reflects the interaction of the phenotype with the environment.

Three other terms have played their part in the past in describing the phenotype. If the effects of the gene are variable then its expressivity can be said to be variable. If the condition controlled by the gene is not always manifested then the gene is said to exhibit incomplete penetrance. If the gene clearly has several phenotypic effects, it is said to be pleiotropic. This term was originally used in connection with the problem of the number of primary products a gene could have, but is useful in the more general sense employed here. All genes are pleiotropic, since our awareness of them depends on our ingenuity in observing and describing the phenotype. Unbanded in *Cepaea* is known from only one effect at present, however, while curly in *Drosophila* is clearly pleiotropic. The effect on the wings is dominant with fairly low penetrance and variable expressivity while the effect on survival is recessive with high penetrance.

Two or more loci may interact in their effect on the genotype. If an allele at one locus masks the effect of an allele at a different locus it is said to be

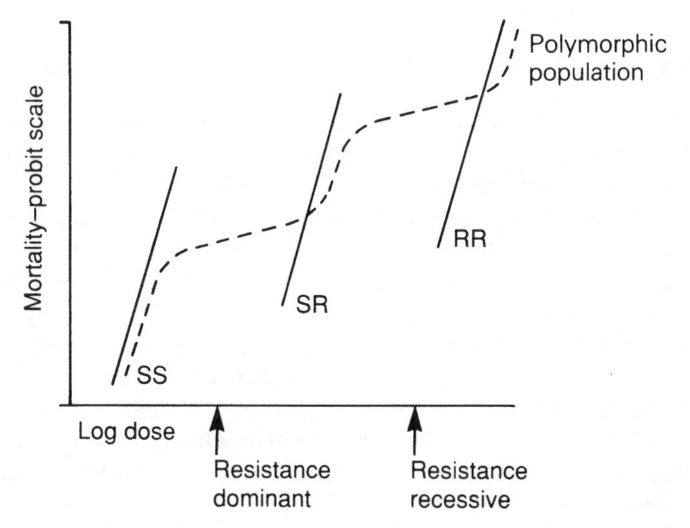

Fig. 5.1 Relation of mortality to dose of insecticide for a population segregating for an insecticide-resistance gene. If mortality transformed to a probit scale is plotted on dose and susceptibility of individuals is normally distributed the result will be an ascending straight line. The diagram shows a characteristic pattern of response when there is a resistance gene conferring intermediate resistance on the heterozygote. In the laboratory, the relative resistance levels can be established by setting up the appropriate tests, but in the field there will be a single mean dose level, and the gene will behave as dominant or recessive depending on what that level is.

epistatic to it (the other allele is hypostatic, although this term is rarely used). Operationally, epistasis is detected in an F_2 cross if there are only three phenotypes among the progeny. In *Cepaea* the allele midbanded is dominant to non-midbanded at one locus and modifies the expression of the five-banded phenotype determined by a second locus. The F_2 cross will produce unbandeds, midbandeds and five-bandeds in the ratio 12:3:1. In the mouse, agouti is an allele dominant to various other coat colour alleles at the agouti locus, while albino is recessive to non-albino at a different locus. In the homozygous albino individual, variation at the agouti locus cannot be detected, and the F_2 cross produces agouti, non-agouti and albino in the ratio 9:3:4. The dominant, recessive relation of alleles at one locus is independent of, and may be in the opposite direction to, the epistasis/hypostasis relation of alleles at different loci.

Genes such as pesticide-resistance genes are recognized directly because of their effect on the viability of their carriers. Others which we recognize through some other trait may also affect the relative fitness of their carriers. It is therefore possible to think of fitness itself as one of the pleiotropic effects of the gene, which may be dominant, recessive, subject to epistatic interaction or variable in penetrance and expressivity. Fitness is commonly treated in this way in population genetics because it is the most direct way of describing the influence of the genotype on change in gene frequency.

5.4 MOLECULAR GENETICS

Genes are not units in the sense first visualized, and one of the most important fields of genetics is concerned with the transfer of information between the DNA sequence and the polypeptide chain. The information is carried from generation to generation as a sequence of triplet codons in a DNA strand. Many, although not all, of the triplets involved code for amino acids, and the message is transcribed by attachment of mRNA.

The code includes three terminator codons which stop the copying and synthesis, and thus fix the length of sequence concerned. Messenger RNA is copied off the DNA through the action of RNA polymerase, and the process starts at RNA polymerase binding and transcription initiating sequences 10–40 base pairs before the point at which transcription begins. These sequences usually consist mostly of adenine and thymine. The gene could therefore be thought of as the length of DNA from the start of the binding section to the transcription terminator, perhaps an average length of about 1000 base pairs or 300 codons. However, this length includes not only the sequences between the initiator section and the beginning of transcription, but also so-called introns, which are parts of the sequence excised before translation. Whatever else they may be found to do, these sequences do not code for amino acids in the final polypeptide chain. If we call this sequence a gene, an operational problem now arises because genes can mutate, recombine and function but difference sequence lengths may be required in each case. The smallest unit which can mutate is the single base. The smallest unit which can recombine is much shorter than the distance from binder to terminator. The functional unit concerned with the production of a given product is essentially the chain we have described, but even here the relationship of gene to its product is not completely defined because post-translational modification of the polypeptide may occur to form the final molecule we associate with the gene. When the gene mutates, the effect may be to cause replacement of one amino acid by another if it is a base change, or a reading frame shift if it is a deletion or addition of a base. The change may affect efficiency but not composition if it is in the promotor region or it may have no outward effect at all if it is in an intron or is a change to the third base to form a synonymous codon.

5.5 GENE ACTION

In order to form the completed organism many genes have to operate together. The numbers quoted range from hundreds in the smallest bacteria such as *Mycoplasma*, to 3000–4000 in the bacterium *E. coli*, 6000–10 000 in *Drosophila* and 15 000–30 000 in man. An essential feature is that they should function at the right time and produce the correct amount of product. We know something about the control processes but the picture is tantalizingly incomplete.

In prokaryotes several operons have now been studied in detail, which consist of the production unit plus the controlling system. In *E. coli*, for example, the tryptophan synthetase (*trp*) operon consists of a site controlling production of a repressor, activated by the presence of tryptophan, a promotor sequence and a sequence coding for enzymes involved in the synthesis of tryptophan. The promotor sets the enzyme synthesis in motion, but is blocked by the product of the repressor. There is, therefore, a feedback system ensuring that enzyme production is switched off when the enzyme has performed the required operation. The *lac* operon which controls breakdown of lactose, functions in the opposite way, in that initiation of the process depends on the presence of lactose. It is tempting to equate the operon with the gene, but in both cases the regulatory site is quite distant from the structural gene region which produces the enzymes. If we do so, the gene is nothing like a continuous sequence, but is spread about the genome. The control systems in eukaryotes are less well established, but enhancer regions are known to be present near to the promotors of functional genes, and whatever the details, the control is by feedback analogous to that in the prokaryote systems.

A genetic system which has been analysed is the production of haemoglobin in man. The oxygen-carrying molecule comprises the haem group wrapped up in two pairs of globin polypeptides. These are produced by genes on chromosomes 11 and 16. On each chromosome there are several sites side by side, each synthesizing a particular globin chain, which have probably evolved by duplication and subsequent divergence of structure. Apart from rarer variants, the average human being has three types of haemoglobin. In adults, most of it is HbA, which consists of two α chains, synthetized on chromosome 16 and two β chains from chromosome 11. There is also some HbA_2, consisting of two α and two β chains. The δ locus is also on chromosome 11. Unborn fetuses have fetal haemoglobin HbF, with two α and two γ chains. The change from fetal to adult haemoglobin takes place at birth. There is therefore no 'gene for haemoglobin'. There are genes for the globins, and these consist of a family of sequences producing similar but distinct products and spread over two chromosomes. The operation of these genes is controlled so that different combinations are employed at different times of life.

Genes which code for a particular product are often called structural genes. If we examine the range of organisms from the most simple to the complex, it is found that the same gene products turn up over and over again, often performing very different functions in different types of organism. The number of different gene products is not enormous: something of the order of 10 000. There is an increase in number from bacteria to eukaryotes, but not by more than two orders of magnitude. In evolution, however, a vast array of diverse forms has come into being, suggesting that much, perhaps most, of evolution has been a process of change in the control of patterns of development rather than the production of new types of component molecules. The total amount of DNA in the genome has increased by more than four orders of magnitude from viruses to the organisms with the largest amounts, which happen to be amphibia and lilies. This 'non-structural' DNA is composed mostly of repetitive sequences which exist in 10-fold to 1000-fold copies. Their function is largely unknown. The coincidence of increase in sophistication of genome control and increase in the fraction of repetitive sequences could be taken to indicate that they are in some way involved in the control process. Some, however, are repeated as a result of unequal crossing over or other mechanisms such as the action of restriction enzymes, which cut out sections of DNA from one place and insert them somewhere else, so that the number of copies of a particular sequence increases. If the duplicating segment is a functioning gene then this process will increase the potential amount of the gene product which can be produced, which could be beneficial or deleterious. If it is a non-coding sequence, however, then the repeats could simply accumulate without any direct benefit accruing, until limited by some disadvantageous effect.

5.6 NON-SEGREGATING CHARACTERS

Another class of inherited traits are those which do not segregate in progeny. They are called continuously varying, non-segregating or biometrical characters. Many of them, such as height, weight, colour or shape of a part of an animal or plant, are truly continuous in that distinctions between classes depend only on the conventions of measurement. Thus, the height of an individual may be measured in centimetres, millimetres or micrometres, depending on the accuracy of the instrument used. In others, such as clutch size, bristle number or ray floret number, variation is by integer steps. These are called meristic variables. The underlying control is the same in each case.

It is often apparent that variation of this type is likely to be of direct importance in evolution. Individuals which differ in fecundity, rate of development or robustness are likely to differ in fitness. These differences were at the root of the controversy which developed at the beginning of the twentieth century over the genetic basis and evolutionary importance of the two classes of trait. Comparison of parents and offspring showed that the continuously varying characters were often inherited. In the nineteenth century, Francis Galton developed the regression coefficient as a measure of the degree of relationship between parents and offspring. Something of the trait was inherited and laws could be formulated relating the slope of the graph of offspring value on parental value to the degree of relationship. A little later, Weldon examined data on one of the classical Mendelian segregating characters, round versus wrinkled shape in peas, and showed that the phenotypic expression of the two classes of pea differs depending on the strain of plants from which the seeds came. Expressivity of a segregating gene was subject to continuous variation, so that rules for regression and relatedness could be applied to all types of inherited character.

In the first two decades of the twentieth century, various pieces of evidence came together to show what kind of genetic control was involved. The Danish geneticist W. Johannsen measured seed weight in French beans and established the variance of this character in an outbred base population. The plants were then inbred. When like individuals are bred, anyone which is homozygous will breed true while a heterozygote will produce 50% homozygotes among its progeny. The fraction of heterozygotes at each locus will halve in each succeeding generation. This must reduce variability if it is controlled by segregating genes. Johannsen found that the variance did indeed go down, rapidly at first and then more slowly until a steady level was reached. The phenotypic variance at the start was composed of at least two parts, a genetic component which could be reduced by inbreeding and a part which could not be altered and was presumably due to environmental effects.

Another thread was developed by examining Mendelian genes lying at different loci but producing the same phenotype. One such example was red versus white grain colour in wheat, for which there are three similar loci. If two loci are homozygous for white, the third can be shown to segregate 1:2:1 for red to pink to white in a cross of two heterozygotes. When two loci are heterozygous the Mendelian expectation is 1:4:6:4:1 for decreasing degrees of redness, while for three loci the expectation is 1:6:15:20:15:6:1. These frequencies form binomial distributions. As the number of loci increases, the distinction between successive categories lessens, provided the effects of alleles at each locus add on to those of the others. With some environmental blurring as well, even two or three loci behaving in this way will produce an apparent continuously varying phenotype. The conditions required are that the effects of alleles at each locus are similar, small, additive and independent. These conditions were met for the grain colour condition in wheat.

Once it is realized that continuously varying characters could be controlled by segregating genes in this way, it is possible to investigate theoretically what the effect would be if some of the assumptions were varied. For example, there may be linkage between loci, some alleles may be dominant, there may be epistatic interaction, etc. R.A. Fisher did so in a very influential paper published in 1918, in the course of which he introduced the use of variance as a measure for analysing quantitative genetics.

Of course, the fact that something is possible does not mean that it is true. The next step is to test the theory by setting up experiments in which changes in mean and variance for the character measured may be predicted from the genetic model. An approach based on work at the turn of the century by the American geneticist E.M. East is shown in Figure 5.2. Starting from a base population obtained by mixing a variety of strains in order to increase genetic variability, lines are selected which are high or low for the character studied. These will form the parental stock for two further crosses. If the trait is inherited the two lines will diverge to different mean values, and their variances will decrease as they become progressively more homozygous. Crossing of individuals from these lines will form an F_1 stock with an intermediate mean value and a low variance, because individuals will tend to be heterozygous. In an F_2 genetic stock, recombination will increase the variance but leave the mean unchanged.

Over the years analysis of all manner of quantitative traits has amply confirmed the multifactorial genetic model. Analysis is based on means, variances and the degree of relationship between relatives, and from these parameters the relative importance of various genetic factors is inferred: number of loci, fraction with additive genes, amount of genotype/environment interaction and so on. Sometimes studies have been continued to identify the locations on chromosomes at which genes affecting the trait are situated. Polymorphic DNA markers are now used to map genes of

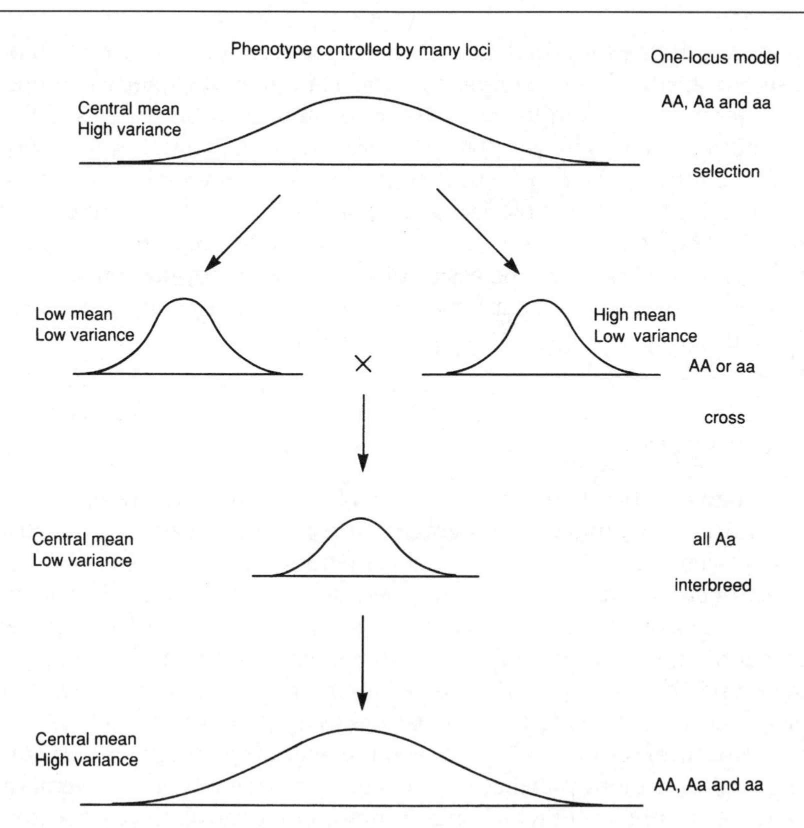

Fig. 5.2 A test of quantitative inheritance. The parallel between mean and variance of phenotypic character and the genetic system assumed to underlie it. Segregation of one locus is shown, but many loci are likely to be involved in most quantitative characters.

major quantitative effect, known as quantitative trait loci or QTLs. The evidence relating genotype to phenotype is indirect, however. It is not usually possible, for example, to say that if we study leaf length in tobacco plants the genes identified are there specifically for the purpose of controlling leaf length. This property might be one of numerous pleiotropic effects, and the number of loci involved could vary from one environment to another. The underlying segregating gene model has only a loose connectedness with the phenotypic variation.

5.7 CONCLUSION

It is clear that our knowledge of genes is very incomplete and, more subtly, that our picture of the gene changes depending on the approach used to study it. Population genetics is about the way selection and other forces modify the inherited characteristics of organisms. Often the models set up start with very simple characterization of the inherited properties in terms of the frequencies of Mendelian segregating alleles. This is done because for many purposes it works very well, and because it is a good scientific method to set up a simple model and then investigate the way in which certain introduced complexities modify the conclusions from it. The use of simple genetic models does not imply a naive approach to inheritance. The assumption of simple population structure as a working method in Chapter 4 does not imply that plants and animals all live that way.

5.8 SUMMARY

Mendelian genetics is founded on a model in which genes are considered to be indivisable units which segregate at reproduction and are redistributed between offspring. This model is operationally sufficient in a wide variety of circumstances. Using it, genetic loci can be described as having alleles with dominant, recessive or codominant effects which segregate independently of, or are linked to, other such loci. When we record the phenotype, this description is usually adequate, especially with the addition of terms such as penetrance, expressivity and epistasis to describe other aspects of the effect of the loci on the phenotype singly or in groups. All genes are pleiotropic, that is, they affect the phenotype in several different ways, only one of which may be apparent to the observer. The dominance of the different pleiotropic effects is not necessarily the same. The inherited component of continuous variation of phenotypic characters is provided by the combined action of genes with small and indistinguishable effects on the character.

At the molecular level, a gene is a linear sequence of DNA made up of an initiator and a stop codon, between which are coding and non-coding regions. At other places in the genome are sequences which determine whether, and by how much, transcription shall take place. The properties of the molecular gene do not necessarily map conveniently on to the Mendelian descriptions. The units of function, mutation and recombination are not equivalent.

Genes interact with others in the genome, so that the effect of any given locus depends on production and timing of activity of other loci. The difference in number of genes between the simplest and the most complex organisms is relatively small; much evolutionary change is the consequence of change in the patterns of interaction rather than of evolution of genes with new products. Mendelian language is usually used in population genetics, but explanations require an awareness of the molecular structure and the effects of gene interaction.

5.9 FURTHER READING

Berg, P. and Singer, M. (1992) *Dealing with Genes. The Language of Heredity.*
Hartl, D.L. and Jones, E.W. (1998) *Genetics: Principles and Analysis.*
Lewin, B. (1997) *Genes VI.*
Watson, J.D, Hopkins, N.H, Roberts, J.W. *et al.* (1987) *The Molecular Biology of the Gene.*
Weaver, R.F. and Hedrick, P.W. (1997) *Genetics.*

Sources of continuous variation 6

6.1 COMPONENTS OF PHENOTYPIC VARIATION

Phenotypes such as those concerned with height or weight often show continuous variation, some of which is genetic in origin while some is environmental. There may also be different expression of the genetic component in different environments (genotype–environment interaction). The general relation of continuously varying phenotypes to the multi-locus genetic control which underlies them is outlined in section 5.6. It is often easy to measure the phenotypic variation but difficult to interpret it. Even where a gene is identified which has a major effect on a quantitative character, such as Mendel's tall/short gene in the pea or Sturtevant's Bar-eye gene in *Drosophila*, there are still numerous genes which exert minor effects on the phenotype. In the same way, environmental effects on quantitative characters are usually manifold and difficult to define. Even in experiments specifically designed to measure genetic and/or environmental effects, there always remains a host of effects which cannot be deciphered individually. These residual effects constitute the so-called error component against which detectable effects are tested for statistical significance.

Because of the manifold and unpredictable nature of the genetic and environmental influences on quantitative characters, it is very unlikely that they will all exert similar effects, so that extreme phenotypes are rare. Intermediate phenotypes will, by contrast, be common since they can result from vast numbers of different combinations of effects. Such considerations explain why quantitative characters usually exhibit normal (bell-shaped) frequency distributions. The shape of the normal curve can be completely defined once two parameters are known: the mean and the standard deviation. The mean determines the region of central tendency of the distribution. The standard deviation (the square root of the variance) measures the spread of the normal curve. Variances have the useful property that they can be partitioned into component parts when they are generated by several different influences. They are therefore the most valuable measure of variability in quantitative biology.

In theory, the deviation of each phenotype from the mean of its population or species can be separated into as many different sections as there are genetic and environmental effects. In practice, this is never possible, not only because the number of effects is unknown but also because any experiment to measure even a large finite number of effects would be impossibly complex. Nevertheless, we can reach some helpful conclusions if we summarize the genetic and environmental effects. Let P_i represent the phenotype of the ith member of a population of N individuals and let us assume that this phenotype is the combination of a general genotypic effect, G_i, and a general environmental effect, E_i, such that

$$P_i = G_i + E_i$$

and therefore

$$\mu_P = \mu_G + \mu_E$$

μ_P, μ_G and μ_E, respectively, representing the mean phenotypic, genotypic and environmental effects. In the standard notation,

$$\sum_{i=1}^{N} P_i = P_1 + P_2 \ldots P_N$$

For tidiness in the following sections, summation from 1 to N will be indicated simply by Σ. The mean (μ_P) of a population of N individuals is given by the expression

$$\mu_P = (1/N)\, \Sigma P_i$$

We can represent the phenotypic variance (V_P) of this population as the average squared deviation about the mean, giving

$$V_P = (1/N)\, \Sigma(P_i - \mu_P)^2$$

This can also be expressed in terms of its genotypic and environmental components, giving

$$V_P = (1/N)\, \Sigma\{(G_i + E_i) - (\mu_G + \mu_E)\}^2.$$

On rearrangement, we have

$$V_P = (1/N)\, \Sigma\{(G_i - \mu_G) + (E_i - \mu_E)\}^2.$$

Expanding the brackets gives us,

$$V_P = (1/N)\, \Sigma(G_i - \mu_G)^2 + (1/N)\, \Sigma\, (E_i - \mu_E)^2 + 2\,(1/N)\, \Sigma(G_i - \mu_G)(E_i - \mu_E).$$

The first two expressions after the equals sign are the genotypic (V_G) and environmental (V_E) variances, respectively. The third expression is the covariance of genotype and environment (W_{GE}). It measures the degree to which genotypic and environmental effects are related. Such a relation is termed the genotype–environment interaction. Thus, in summary, we have

$V_P =$	V_G	$+ \; V_E$	$+ \; 2W_{GE}$
Phenotypic variance	Genotypic variance	Environmental variance	Genotype–environment interaction

The genotypic variance may, in turn, be split into additive, dominance and epistatic components, giving

$V_G =$	V_A	$+ \; V_D$	$+ \; V_I$
Genotypic variance	Additive variance	Dominance variance (allelic interactions)	Epistatic variance (non-allelic interactions)

Additive effects represent the direct result of substituting one allele for another, as when we compare the phenotypes (symbolized by square brackets) of two homozygotes differing only at a single locus ([AA] versus [aa]). As we shall see, these are by far the most important in determining the potential response to selection. Dominance effects measure the degree to which the phenotype of a heterozygote differs from the mean of the two corresponding homozygotes ([Aa] versus [AA + aa]/2). Such differences reflect interaction between allelic effects in the heterozygote. Epistatic effects reflect interactions between the effects of non-allelic genes. They are numerous and complex but often signify situations where the effect of one gene on part of a biosynthetic sequence is subject to the effect of another on an earlier stage of the biosynthesis.

6.2 POLYGENIC MODELS

6.2.1 PREMISES

Let us construct a simple model of a polygenic system, consisting of two gene loci each with two alleles of equal minor effect (a diallelic system). We shall adopt the convention whereby the allele of increasing effect on the phenotype is designated by a capital letter and that of decreasing effect, by a letter in lower case. The notation does not involve any implicit assumption of dominance. We shall assume that non-allelic genes do not interact, in other words there is no epistacy, and that the loci are unlinked.

In addition, we will be concerned with models of theoretical populations each of N individuals and therefore with formulae for the parameters themselves, rather than estimates of those parameters obtained from samples of n individuals. In practice, the only difference between the formulae for samples and populations is that the formula for a sample variance includes a denominator of $n - 1$ rather than N, as is the case for a population variance. The reason for this is that a sample variance has only $n - 1$ degrees of freedom, one degree of freedom being lost because the sample variance is based on its own sample estimate of the mean and not on the true population mean. This has already been discussed in section 4.6. At this point, however, it is useful to introduce a working formula which is easier to compute but not so obvious as its analytical counterpart. Thus, the variance (V) of a series of phenotypes P_1, $P_2 \ldots P_N$ with a mean of μ can be represented as either

$$V = (1/N)\,\Sigma(P_i - \mu)^2 \qquad \text{or} \qquad V = (1/N)\,\Sigma P_i^2 - \mu^2$$

Analytical formula — Working formula

the working formula for the variance being employed in our models. When we use proportions, we effectively set $N = 1$.

6.2.2 ADDITIVE EFFECTS

If **A** is given a phenotypic effect of + 0.5 while its allele **a** has an effect of – 0.5, three diploid genotypes are possible with three distinct phenotypes.

These are:

	AA	**Aa**	**aa**
Genotype	**AA**	**Aa**	**aa**
Phenotype	[+ 1]	[0]	[– 1]

The phenotypic effects of **A** and **a** are purely additive, not only in the homozygotes where only one type of allele is present but also in the heterozygote where opposing effects of the different alleles cancel out on addition. Such simple additivity represents the great majority of minor effects within natural polygenic systems.

Next, we will introduce a second purely additive locus **B**/**b**, **B** being equivalent in effect to **A** and **b** being equivalent to **a**. There are now two distinct categories of double homozygote which could form pure-breeding inbred lines: those showing **association** of phenotypic effects and hence extreme phenotypes (**aabb**: [– 2], **AABB**: [+ 2]) and those showing **dispersion** of phenotypic effects and hence intermediate phenotypes (**aaBB**: [0], **AAbb**: [0]). These phenomena of association and dispersion are of great importance in evolution and in commercial breeding. Two inbred lines may have similar phenotypes due to dispersion and yet radically different genotypes. Much valuable additive genetic variation may therefore remain concealed until exposed in any future F_2 progeny. The contribution of additive genetic effects to the F_2 generation can be investigated, using our model.

Two inbred lines of extreme phenotype are crossed to produce an F_1. Convention has it that the inbred line of larger phenotype is designated P_1 and that of smaller phenotype, P_2.

$$\begin{array}{ccc} \textit{Cross} & P_1\text{: } \mathbf{AABB} \times P_2\text{: } \mathbf{aabb} \\ & [+2] \quad\quad [-2] \\ & \downarrow \\ & F_1\text{:}\mathbf{AaBb} \\ & [0] \end{array}$$

Selfing the F_1 individuals and raising a total of 16 offspring, we expect to obtain nine different genotypes in the F_2, representing five different phenotypes. Details of these are given under Model 1 in Table 6.1, along with the sums of X and X^2 from which the mean and variance have been calculated. The mean of zero and variance of unity demonstrate that, while

Table 6.1 Comparison of the influences of dominance and differences in gene frequency on means and variances in a simple polygenic model of F_2 progeny where each gene contributes a minor effect (±0.5) to the phenotype (X_i)

	Model 1		Model 2		Model 3		Model 4		Model 5	
X_i	Dominance	none	Dominance	none	Dominance	A	Dominance	A, B	Dominance	A, b
	p(A)	0.50	p(A)	0.75	p(A)	0.50	p(A)	0.50	p(A)	0.50
	p(B)	0.50	p(B)	0.75	p(B)	0.50	p(B)	0.50	p(B)	0.50
	Genotype	*Freq.*	*Genotype*	*Freq.*	*Genotype*	*Freq.*	*Genotype*	*Freq.*	*Genotype*	*Freq.*
[−2]	aabb	1	aabb	1	aabb	1	aabb	1	aabb	1
									aaBb	2
										3
[−1]	Aabb	2	Aabb	6	aaBb	2				
	aaBb	2	aaBb	6						
		4		12						
[0]	AAbb	1	AAbb	9	AAbb	1	AAbb	1	AAbb	1
	aaBB	1	aaBB	9	aaBB	1	aaBB	1	aaBB	1
	AaBb	4	AaBb	36	Aabb	2	Aabb	2	Aabb	2
		6		54		4	aaBb	2	AABb	2
								6	AaBb	4
										10
[+1]	AaBB	2	AaBB	54	AaBb	4				
	AABb	2	AABb	54	AABb	2				
		4		108		6				
[+2]	AABB	1	AABB	81	AABB	1	AABB	1	AABB	1
					AaBB	2	AaBB	2	AaBB	2
						3	AABb	2		3
							AABb	4		
								9		
N		16		256		16		16		16
ΣX		0		256		8		16		0
ΣX^2		16		448		24		40		24
μ		0.00		1.00		0.50		1.00		0.00
V		1.00		0.75		1.25		1.50		1.50

additive effects may cancel each other out in their contribution to the mean, they do not oppose each other in their contribution to the variance. This result provides a yardstick, against which we may assess the effects of increasing numbers of loci, variation in gene frequency and non-additivity.

6.2.3 EFFECTS OF INCREASING NUMBERS OF LOCI

Given a single diallelic locus such that **A** has an effect of + 0.5 and **a** an effect of – 0.5, the mean of the F_2 will be zero,

$$\mu = \tfrac{1}{4}[-1] + \tfrac{1}{2}[0] + \tfrac{1}{4}[+1] = 0$$

while the variance will be half

$$V = (\tfrac{1}{4}[-1]^2 + \tfrac{1}{2}[0]^2 + \tfrac{1}{4}[+1]^2) - 0^2 = 0.5.$$

Comparing this with the previous result where there were two such loci (Table 6.1), we see that the mean has remained unaltered but the variance has increased from 0.5 with a single locus to 1.0 with two loci. Each additional locus of the same effect makes the same additive contribution to the variance. An equivalent system based on three loci will have a variance of 1.5, one based on four loci, a variance of 2.0, and so on. In general, k loci of this sort would produce a mean of zero but a variance of $k/2$.

Alternatively, we may envisage a model where, as the number of loci increases, the phenotypic effect at each locus decreases so that the variance remains constant. For example, we might wish to maintain a mean of zero and a variance of unity. Under these circumstances, we have

$$V = (1/2k)\sum_{i=1}^{2k} d_i^2 - 0^2 = 1 \text{ so that } \sum_{i=1}^{2k} d_i^2 = 2k$$

where d_i represents the additive effect at each locus. Assuming these effects to be constant ($d = d_i$, for all values of i), then

$$\sum_{i=1}^{2k} d_i^2 = \left(\sum_{i=1}^{2k} d_i\right)^2 = (2kd)^2 = 2k \text{ so that } 2kd = \sqrt{(2k)} \text{ and } d = 1/\sqrt{(2k)}.$$

A similar argument shows that haploid systems can have a mean of zero and a variance of unity so long as $d = 1/\sqrt{k}$. Substitution into these expressions shows how the phenotypic contributions of each locus decrease as the number of loci increases (see Table 6.2).

6.2.4 EFFECTS OF VARIATION IN GENE FREQUENCY

The term gene frequency always refers to the relative frequency of an allele compared with others at the same locus. Given two alleles **A** and **a**, of respective frequencies p and q, the three possible genotypes formed under conditions of random mating will have frequencies determined by the terms produced by expansion of the binomial $(p + q)^2$ (see Chapter 7).

These are:	Genotype	**AA**	**Aa**	**aa**
	Frequency	p^2	$2pq$	q^2
	Phenotype	[+ 1]	[0]	[– 1]

Given k non-allelic loci each of two alleles similar in frequency and effect to **A** and **a**, there will be $2k + 1$ possible phenotypes in frequencies determined by expansion of the binomial $(p + q)^2k$. Under such circumstances, the mean phenotype will be

$$\mu = k(p^2[+1] + 2pq[0] + q^2[-1])$$
$$= k(p[+1] + q[-1]),$$

as expected.

The variance in phenotype is given by the expression

$$V = k\{(p^2[+1]^2 + 2pq[0]^2 + q^2[-1]^2) - \mu^2\} = 2kpq.$$

Varying p in the expression will show that V has a maximum value when $p = ½$.

Evidently, phenotypic variances are maximal when gene frequencies are equal, and as they become progressively more unequal, so the variance declines. Obviously when $p = 1$ so that there is no alternative allele, the variance will be zero. Such considerations are important both in commercial breeding programmes and in evolution.

Effects of particular gene frequencies on phenotypic means and variances can be derived from the equations of the previous section. Thus, if three-quarters of gametes contain **A** and the remaining quarter contain **a** and a similar situation prevails for a second locus **B**/**b**, then $p = 0.75$ and $k = 2$. We then have

Table 6.2 Average genic effects consistent with a mean of zero and variance of unity in polygenic systems with different numbers of loci

Number of loci k	Diploids Effect/locus $1/\sqrt{2k}$	Haploids Effect/locus $1/\sqrt{k}$
1	0.707	1.000
2	0.500	0.707
3	0.408	0.577
4	0.354	0.500
10	0.224	0.316
50	0.100	0.141
100	0.071	0.100

Table 6.3 Gametes from F_1 individuals in which the two alleles at each of two loci differ in frequency by a factor of three to one

	Locus 1	
	A (3/4)	**a** (1/4)
Locus 2	F_1 gametes	
B (3/4)	**AB** (9/16)	**aB** (3/16)
b (1/4)	**Ab** (3/16)	**ab** (1/16)

$$\mu = k\,(2p - 1) = 2\,(1.5 - 1) = 1.00$$

and

$$V = 2kpq = 4\,(0.75)\,(0.25) = 0.75.$$

We can achieve the same result from first principles by substituting directly into the model. To do this, we assume that the four types of gamete produced by the F_1 generation occur in proportions determined solely by the relevant gene frequencies. The outcome is represented in tabular form in Table 6.3.

By combining the gametes at random, we obtain the distribution of genotypes and phenotypes amongst 256 F_2 individuals detailed under Model 2 in Table 6.1. From these data we obtain

$$\mu = 256/256 = 1$$

$$V = 448/256 - 1^2 = 0.75,$$

as we found before.

Evidently, the effect of unequal gene frequency is to move the mean in the direction of the phenotypic effects of the common alleles (i.e. in this case, from zero to unity) but to decrease the variance (from unity to three-quarters). Had **b** been more common than **B** and as common as **A**, then the effects of unequal gene frequency at the two loci (**A/a** and **B/b**) would have cancelled each other out to produce an unchanged mean of zero (by comparison with the situation of equal gene frequencies) but the variance would still have decreased to three-quarters.

6.2.5 EFFECTS OF DOMINANCE

When **A** is completely dominant over **a**, the genotype of the heterozygote is indistinguishable from that of a homozygote for the dominant allele.

These are:	Genotype	**AA**	**Aa**	**aa**
	Phenotype	[+ 1]	[+ 1]	[– 1]

Here, although the effects of the two alleles prove additive when homozygotes are compared, they interact in the heterozygote. Dominance is therefore the expression of allelic interaction.

The effects of complete dominance may be explored using the simple model. We shall assume three different scenarios. They are (Model 3 of Table 6.1) dominance of **A** over **a** but no dominance at locus **Bb**, (Model 4) dominance of **A** over **a** and of **B** over **b** and (Model 5) dominance of **A** over **a** but dominance of **b** over **B**. The situation in Model 4 is known as **reinforcing dominance**, because the dominance effects are in the same direction, whereas Model 5 is an example of **opposing dominance**. The genotypes will be present in the same frequencies as in Model 1, with equal gene frequencies, and the only phenotypes affected by dominance will be those of the heterozygotes. Thus the F_1 will have a phenotype of [+ 2], rather than [0] in the absence of dominance.

Comparison of Models 3, 4 and 5 in Table 6.1 illustrates that the effect of dominance is to move the mean in the direction of the dominant phenotypic effects and to increase the variance. Two important results emerge from this comparison. First, we see that opposing dominance effects can cancel each other out and hence make no net contribution to the mean. Means are therefore unreliable estimators of dominance. Second, the dominance effect at each locus makes a distinct contribution to the variance, regardless of the direction of dominance. The size of the dominance effect is therefore measurable, at least in theory.

6.3 HERITABILITY AND RESPONSE TO SELECTION

Selection can only change the composition of successive generations if the variation on which it acts has a genetic component. This is as true of natural selection acting on wild populations as it is of artificial selection practised by plant and animal breeders. We can examine relationships between genetic variance and selection potential by studying heritabilities of continuously varying characters.

The degree of genetic determination involved in the phenotypic variation exhibited by a continuously varying character is most simply summarized as the proportion V_G/V_P. This is known as the **heritability in the broad sense**. It is normally given the symbol H^2 (or more recently h^2_B). The more useful measure is the **heritability in the strict sense**, represented by

$$h^2 = V_A/V_P$$

The term heritability is misleading, but has become established in the literature. It does not tell us whether or not a particular characteristic is inherited but does provide a measure of the extent to which variation in that character reflects differences in genotype. We may think of it as describing the degree to which a phenotypic variance is heritable. A character with low heritability may be so important biologically that any genetic variation has been eliminated by strong natural selection, leaving only those environmental effects which are beyond the organism's control (Table 6.4). An obvious example is conception rate in large herbivores such as Friesian cattle. Failure to conceive results in failure to reproduce, yet over-conception produces too many fetuses which will not only threaten each other's survival but also that of the mother. Genetic variation in conception rate has therefore been minimized by natural selection. Very little variation remains and this is almost all environmental, so that the heritability is low. By contrast, although fingerprint ridges in man may well have a function, variation in the patterns is of little importance and a great deal of genetic variability is tolerated. Comparison of heritabilities in a range of domesticated animals and cultivated crop plants shows that reproductive characters tend to have phenotypic variances with the lowest proportions of additive effects (Table 6.4).

The strength of selective forces operating on a continuously varying character is measured as the selection differential (S). This is the difference

Table 6.4 Percentage estimates of heritability obtained from a variety of sources, notably Falconer and Mackay (1996)

Livestock		Crops	
Cattle, Friesian		*Maize*	
White spotting	95	Husk number	47
Butterfat content	60	Cob weight	32
Milk yield	30	Corn weight	26
Conception rate	1	Total yield	24
Pigs		*Spring wheat*, Timgalen × Sonora	
Back fat	55	Height	72
Body length	50	Kernel weight	56
Wt at 180 days	30	Heading date	48
Litter size	15		
Poultry, White Leghorn		*Winter wheat*, Centurk × Bezostaia	
Egg weight	60	Heading date	100
Age at first laying	50	Kernel weight	65
Egg production	30	Height	64
Body weight	20	Tiller number	36
Viability	10		
Sheep, Merino			
Wool length	55		
Fleece weight	40		
Body weight	35		

between the means of selected parents ($\bar{P}_S$) and the mean of all the parents ($\bar{P}$), referred to in section 4.7.1 as threshold selection. Selection differentials can be expressed in absolute units (such as mg or cm) or they can be standardized as relative units of phenotypic standard deviation (s_P). Standardized selection differentials are inversely related to the proportion of parents selected. They therefore represent a measure of selection intensity (i), comparable to the I of section 4.7.

Where phenotypic variation has a polygenic component, offspring will tend to resemble their parents. Such relations can be investigated graphically, by plotting characteristics of offspring against their mean parental values, known as midparents (Figure 6.1). The line of best fit between the points obtained will pass through the central point determined by the mean of all the parents and that of all the offspring. The slope or regression coefficient (b_{PO}) of this line is equal to the parent–offspring covariance (W_{PO}) divided by the variance of the parents (V_P). The parent–offspring covariance is known to measure the additive component (V_A), so the regression coefficient provides a direct estimate of heritability in the strict sense.

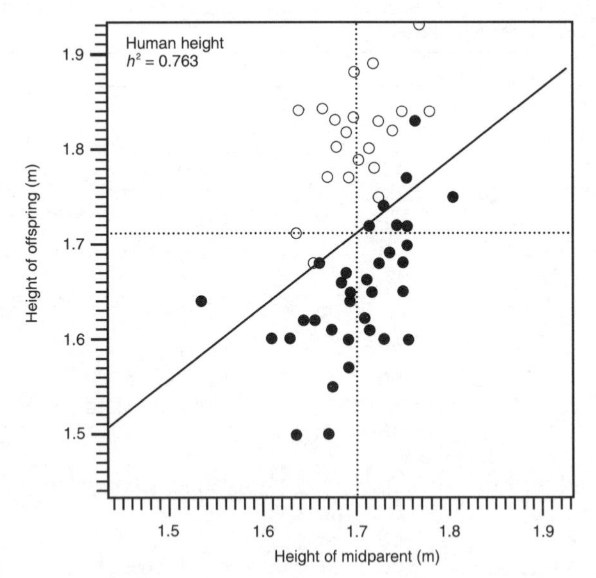

Fig. 6.1 Estimation of heritability using parent–offspring regression for heights of a class of university students who were asked to measure their parents. Height of offspring is plotted on midparental value. Closed circles: females; open circles: males. The heritability is that of all offspring on midparent, estimated as the regression coefficient, which is highly significant ($P < 0.01$). Regression analysis minimizes variation along the Y-axis (offspring) but not the X-axis (midparents). For this reason, it does not always give a good visual fit. Because men tend to be taller than women, families with sons are above the regression line whereas those with daughters are below it. Estimates could be made for each sex separately on midparent, or sons could be plotted on fathers and daughters on mothers. In the case of offspring on a single parent, the heritability is twice the regression coefficient.

The response to selection (R) is measured as the difference between the mean of the offspring of selected parents ($\bar{O}_S$) and the mean of all the offspring ($\bar{O}$). The response can be predicted as the product of the selection differential (S) and the heritability in the strict sense (h^2). Thus we have

$$R = h^2 S$$

In addition to providing an estimate of heritability, parent–offspring regression neatly summarizes the interaction between selection and components of phenotypic variance. We have already seen that the response to selection is determined by the proportion of the phenotypic variance ascribable to additive genetic effects. However multifaceted the environment may be, we can distinguish two conflicting sets of environmental factors: those that make up natural selection which depend on genetic variation, and those which constitute the non-heritable component of developmental variation. The greater the developmental component of phenotypic variance, whether due to genotype–environment interaction or to purely environmental effects, the lower the heritability and the lower the response to selection. The parent–offspring regression line is like the dial on a meter, its gradient reflecting the proportion of additive variance and the potential response to selection. Increased environmental input to developmental variation causes the dial to dip and suppresses the potential response to selective forces in the environment.

6.4 GENETIC AND ENVIRONMENTAL EFFECTS AND THEIR INTERACTION

Additive and interactive effects can be distinguished by specialized statistical procedures such as factorial analysis of variance. At its simplest, factorial analysis of variance separates out the effects of two factors and their interaction. For example, the ith individual of the jth genotype reared in the kth environment might have a phenotype ($P_{i,j,k}$) with the following composition.

$$P_{i,j,k} = \mu + G_j + E_k + GE_{j,k} + r$$

G_j representing the deviation from the mean due to genotype j, E_k representing the deviation due to environment k and $GE_{j,k}$ measuring the effect of the interaction between genotype j and environment k. The residual item, r, represents all those genetic and environmental effects which have not been accounted for in the experiment. The more factors in the analysis, whether genetic or environmental, the more interactions are possible and the more complex these interactions become.

We can see how the procedure works by examining some unpublished data obtained by Angus MacEwan on clones of *Poa alpina*, grown under controlled environmental conditions. *Poa alpina* is a grass of high altitude and latitude in the northern hemisphere which often reproduces parthenogenetically to form seed of identical genotype to that of the mother plant. Numerous seed-producing clones have been identified and many of these have distinctive chromosome numbers resulting from unequal gains or losses of chromosomes at some stage in their histories. The great advantage of parthenogenetic seed is that the performance of cohorts of nearly identical progenies can be compared in contrasting and highly regulated environmental conditions.

Seed from four distinct clones was employed: SLP1 ($2n = 33$) from a disturbed habitat at 300 m in Abisko (Swedish Lapland), SLP2 ($2n = 44$) from an alpine meadow at 800 m on Mount Njulla (Swedish Lapland), YORK ($2n = 39$) from the summit at 300 m on Ingleborough Hill (North Yorkshire, UK) and LOFT ($2n = 38$) from a roadside verge at 50 m near Reine (Lofoten Islands, Norway). Ten similar-sized seedlings from each of the four clones were cultivated under eight environmental conditions, representing all possible combinations of day length, moisture and temperature. There were two levels of each environmental factor; details of these and the resulting factor means are given in Table 6.5.

The analysis of variance (Table 6.6) confirmed that each of the environmental factors had a significant effect on plant growth, as did the genotypic factor. Three orders of interaction were possible: six first-order interactions (between two factors), four second-order interactions (between three factors) and one third-order interaction (between four factors). Three of the first-order interactions and one of the second-order interactions are purely environmental. The remaining seven interactions all involve some aspect of genotype interaction and four of these are significant.

First-order interactions are the simplest to interpret, as illustrated by inspection of that between genotype and day length (Table 6.7). We see

Table 6.5 Genotypic and environmental factor means for blade length in the grass *Poa alpina*

Genotype	Blade length (cm)	Environment	Blade length (cm)
Clone		*Day length*	
SLP1	9.43	Short (S: 8 h)	11.71
SLP2	10.62	Long (L: 16 h)	11.24
YORK	13.80		
LOFT	10.62	*Moisture regime*	
		Wet (W: watered every 2 days)	11.12
		Dry (D: watered every 4 days)	11.83
		Temperature	
		Cool (C: 5–10°C)	8.39
		Temperate (T: 15–20°C)	14.56

Table 6.6 Factorial analysis of variance of blade length in *Poa alpina*. df, degrees of freedom; *P*, probability; n.s., not significant

Source	df	Mean square	*F*-ratio	*P*
Main effects				
Genotypes (GS)	3	284.30	69.68	<0.001
Day lengths (DL)	1	17.86	4.38	<0.05
Moisture regimes (MR)	1	40.61	9.95	<0.01
Temperatures (TS)	1	3057.00	749.00	<0.001
First-order interactions				
GS × DL	3	48.36	11.46	<0.001
GS × MR	3	3.22	0.79	n.s.
GS × TS	3	43.31	10.60	<0.001
DL × MR	1	89.68	21.98	<0.001
DL × TS	1	396.90	97.27	<0.001
MR × TS	1	369.90	90.66	<0.001
Second-order interactions				
GS × DL × MR	3	6.65	1.60	n.s.
GS × DL × TS	3	15.03	3.68	<0.05
GS × MR × TS	3	1.66	0.41	n.s.
DL × MR × TS	1	130.80	32.06	<0.001
Third-order interaction				
GS × DL × MR × TS	3	15.60	3.82	<0.05
Residual variation	160	4.08		
Total	319			

that, although each clone produces a different average blade length under short- and long-day conditions, the direction of the effect of day length varies according to the clone under consideration. The value of the mean expected for each combination of clone and day length can be predicted from the main effects. For the combination of LOFT in long days, these are the deviations of the LOFT mean, μ[LOFT]), and the long-day mean, μ[L], from the grand mean, μ. From Table 6.7 we have

Ex[LOFT, L]	$= \mu$	$+ \ \mu[\text{LOFT}] - \mu$	$+ \ \mu[\text{L}] - \mu.$
Expected mean	Grand mean	LOFT effect	Long-day effect

$$= 11.48 \quad + (10.62 - 11.48) \ + (11.24 - 11.48)$$
$$= 10.38.$$

This, apart from a slight rounding error, is the value given in Table 6.7. Although we have successfully predicted that this clone will perform less well under long rather than short days, there none the less remains a small discrepancy between the observed and expected means. It is this discrepancy which reflects genotype–environment interaction. Because such

Table 6.7 First-order interaction of genotype and day length, on mean blade length (cm) in *Poa alpina*

Clone		Blade length (cm) Short days	Blade length (cm) Long days	Genotypic mean
SLP1	Observed	9.49	9.38	9.43
	Expected	9.67	9.20	
	Interaction	–0.18	+0.18	
SLP2	Observed	12.69	11.42	12.05
	Expected	12.29	11.82	
	Interaction	+0.40	–0.40	
YORK	Observed	13.60	14.01	13.80
	Expected	14.04	13.57	
	Interaction	–0.44	+0.44	
LOFT	Observed	11.08	10.17	10.62
	Expected	10.86	10.39	
	Interaction	+0.22	–0.22	
Environmental mean		11.71	11.24	Grand mean 11.48

interactions make no net contribution to the main genotypic or environmental effects, they cancel each other out within any row or column of a table of interactions.

Second- or third-order interactions are obviously more complicated than first-order interactions but the principle remains the same, the only difference being that lower-order interactions are included with main effects in the derivation of expected means. When we are looking at higher-order genotype–environment interactions, we are less concerned with individual interaction items than with the general environmental sensitivity of a particular genotype. Such environmental sensitivity can be examined by looking at performance of a clone across a range of environmental means. Performance is assessed as the mean value of the clone in a particular environment, while each environment is represented by the mean of all clones in that environment.

The significant third-order interaction in Table 6.6 indicates that the clones differ in environmental sensitivity. The two clones from Abisko have similar, relatively low, environmental sensitivity despite differences in chromosome number and original habitat. The clones from Ingleborough and the Lofoten Islands are also of similar sensitivity but both are more environmentally sensitive than those from Abisko. In the tobacco species *Nicotiana rustica*, which has been the subject of a great deal of experimental breeding, inbred lines show greater environmental sensitivity than their F_1 hybrids. Such results suggest that the phenotypes of homozygotes are more sensitive to environmental change than those of heterozygotes.

6.5 CONCLUSION

As indicated in section 5.6, the study of statistical methods for analysing quantitative variation was begun in the nineteenth century. Darwin's cousin Francis Galton developed regression analysis. Karl Pearson and Weldon studied correlation, R.A. Fisher was responsible for the analysis of variance technique. These methods have since become of major significance in many fields of science and technology, much as the theory of probability was developed in the seventeenth century by Fermat and Pascal in response to queries from gambling French aristocrats but has now become a mainstay of applied mathematics. The *impasse* between the biometricians and the rediscoverers of Mendel was solved by Johannsen and Nilsson-Ehle in Europe and by East in North America. These biologists showed that Mendel's methods could be applied, with limitations, to the study of continuous variation. Inherited variation was seen to be controlled by large numbers of genes, each of minor effect, constituting polygenic systems. Specialized statistical procedures were then developed, for examining the properties of polygenic systems: additivity, dominance and interaction. Again a small number of biologists made significant contributions, notably Mather, Jinks and Alan Robertson.

In all this activity, the traditional approach of making inferences from phenotypes was employed. People were less concerned with the genes themselves than with their effects. There had been some idea that 'polygenes' were in some way different from 'major genes' but this concept was later abandoned. Some genes do indeed have major effects on continuous variation and these can be identified by their distorting effects on quantitative frequency distributions or on relations between parents and offspring. Such genes are known as quantitative trait loci (QTLs) and their location on particular segments of chromosome can be investigated with the aid of molecular markers. Studies of QTLs are of particular promise in agriculture and medicine.

With the exception of QTLs, as we have seen, the number and location of genes influencing continuous variation are not only impossible to determine but also of no great importance to our understanding of the principal properties of that variation. In this respect, quantitative genetics now represents one of the few areas of genetics which is unlikely to be revolutionized by advances in molecular techniques. We need to be able to distinguish genetic from environmental effects and additive effects from interactions; statistical procedures are more likely to help us in this endeavour than molecular techniques.

6.6 SUMMARY

Continuous variation results from large numbers of minor genetic and environmental effects which result in normal distributions where intermediate types are common but extremes are rare. Phenotypic variance can be partitioned into three broad categories: genetic variance, environmental variance and genotype–environment covariance. Polygenic systems consist of large numbers of genes, each of minor phenotypic effect. The greater the number of genes in the system, the greater the variance if the effect of each gene remains constant, or the smaller the genic effect if the variance stays the same. Variances are maximal when gene frequencies are equal. Additive and dominance effects make separate contributions to the phenotypic variance. The proportion of additive genetic effects is known as heritability in the strict sense. This determines the response to selection. Environmental factors can be broadly categorized into those which constitute selection and those which contribute to developmental variation. The greater the developmental variation, the lower the response to selection. Interaction between genetic and environmental effects can be investigated by factorial analysis of variance. While it may often be useful to itemise first-order interactions, more complex higher-order interactions are best studied as the sensitivity of particular genotypes to a spectrum of environmental conditions.

6.7 FURTHER READING

Falconer, D.S. and Mackay, T.F.C. (1996) *Introduction to Quantitative Genetics.*
Kearsey, M.J and Pooni, H.S. (1996) *The Genetical Analysis of Quantitative Traits.*
Mather, K. and Jinks, J.L. (1971) *Biometrical Genetics.*
Roff, D.A. (1997) *Evolutionary Quantitative Genetics.*

Gene frequency 7

7.1 DYNAMICS OF GENE FREQUENCY CHANGE

Both discontinuous and continuously varying characters appear to be influenced by a system which, at one level of analysis, operates as a set of segregating genes which may mutate and recombine. The simplest approach to studying the dynamics of change in the hereditary material is, therefore, to see what happens at individual loci to change allele frequency. When that has been explored a more complex system can be developed.

7.2 THE HARDY–WEINBERG EQUILIBRIUM

In a diploid organism, each individual carries a pair of the alleles representing each locus. The population may contain several different alleles. What sort of pattern should we expect for such a locus? The usual practice is to define the Hardy–Weinberg equilibrium conditions, published independently by G.H. Hardy and W. Weinberg in 1908. The statement can be written as follows.

Genetic system:		*Population:*		*Result:*
For an autosomal locus	with	random mating, infinite size, no mutation, no selection	there will be	(a) constant gene frequency, and (b) stable genotype frequency (c) in one generation

The fact that the above conditions hold may be demonstrated in the following way for two autosomal alleles. Consider two alleles, which may be designated A and A′ so as to make no implicit assumption about dominance. Let the frequency of alleles A and A′ be p and q ($p + q = 1$). Then start with any genotype frequencies s, t and u for AA, AA′ and A′A′, subject only to the condition that $s + t + u$ add to unity. Then whatever the genotype values, $p = s + \frac{1}{2}t$ and $q = u + \frac{1}{2}t$. When there is random mating we can represent the frequencies of the various matings as

	males		AA	AA′	A′A′
			s	t	u
females	AA	s	s^2	st	su
	AA′	t	st	t^2	tu
	A′A′	u	su	tu	u^2

These produce progeny in the following frequencies.

mating	frequency	frequency in progeny AA	AA′	A′A′
AA × AA	s^2	s^2	–	–
AA × AA′	st	$\frac{1}{2}st$	$\frac{1}{2}st$	–
AA′ × AA	st	$\frac{1}{2}st$	$\frac{1}{2}st$	
AA × A′A′	su	–	su	–
A′A′ × AA	su	–	su	–
AA′ × AA′	t^2	$\frac{1}{4}t^2$	$\frac{1}{2}t^2$	$\frac{1}{4}t^2$
AA′ × A′A′	tu	–	$\frac{1}{2}tu$	$\frac{1}{2}tu$
A′A′ × AA′	tu		$\frac{1}{2}tu$	$\frac{1}{2}tu$
A′A′ × A′A′	u^2	–	–	u^2
sum	1	$s^2 + st + \frac{1}{4}t^2$	$st + 2su + \frac{1}{2}t^2 + tu$	$u^2 + tu + \frac{1}{4}t^2$
	=	$(s + \frac{1}{2}t)^2$	$2(s + \frac{1}{2}t)(u + \frac{1}{2}t)$	$(u + \frac{1}{2}t)^2$
	=	p^2	$2pq$	q^2
	=		$(p + q)^2$	

The gene frequency has been unchanged by this round of mating and segregation, and a particular genotype frequency pattern has been reached. The next step would be to go round again, substituting p^2, $2pq$ and q^2 for s, t and u. The genotype frequency will remain unchanged, and it will always return to these frequencies if disturbed from them. The equilibrium is therefore stable. A similar demonstration may be given for a population with more than two alleles in it. The pattern of genotype frequencies at different gene frequencies for the two-allele case is shown in Figure 7.1.

If the gene is sex-linked then the equilibrium is different. In man, where the locus for red–green colour blindness is on the X chromosome, and males are heterogametic, some data collected over a number of years in a

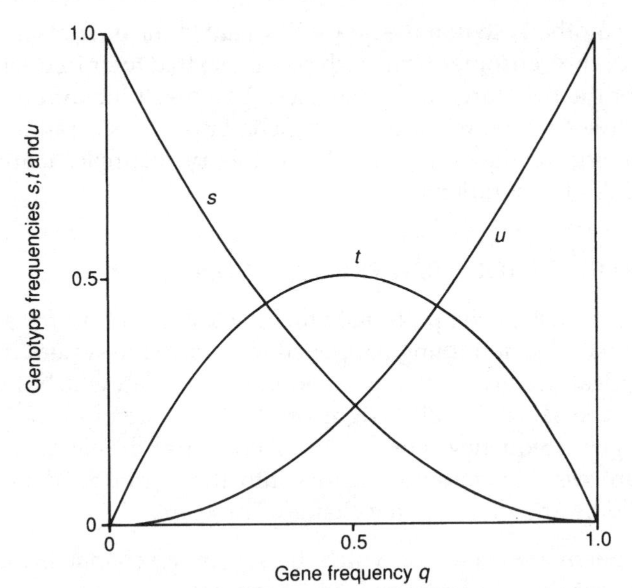

Fig. 7.1 Genotype frequencies for two autosomal alleles at Hardy–Weinberg equilibrium for different gene frequencies. At $q = 0.5$, the two homozygotes have frequencies of 0.25 and the heterozygotes are at their maximum of 0.5.

class practical are shown in Table 7.1. The difference in frequency between the sexes occurs because colour blindness is recessive, so that only homozygotes are detected in females. The frequencies in males are the gametic frequencies since each X chromosome carrying the locus is paired with an inert Y chromosome. In males the estimated frequency of colour blind alleles is $q_m = 34/571 = 0.060$. In females, the colour blind individuals must all be homozygotes, represented by $q_f^2 = 2/613$. The estimated gene frequency is the square root of this quantity, or 0.057. The two frequencies are very close, and the overall frequency estimated for a population with equal sex ratio is $\frac{1}{3}q_m + \frac{2}{3}q_f = 0.058$, since two-thirds of the X chromosomes in the population are carried by females and one-third by males.

But what happens if the frequencies in the two sexes are not close to each other? In each generation male offspring get all their X chromosomes

Table 7.1 Numbers of colour blind and non-colour blind individuals among biology students in Manchester over several years. Scoring is based on ability to discriminate patterns using standard testing charts

	Non-colour blind (no.)	Colour blind (no.)	Total (no.)
Males	537	34	571
Females	611	2	613

from their mothers, so that the new q_m is equal to the old q_f. Daughters get half their X chromosomes from each parent, so that their frequency is half the sum of the two parental frequencies. As a result the difference in frequency between the sexes decreases by a half in each generation, so that q_m and q_f converge on the mean value, but in theory only reach it after an infinite number of generations.

7.3 FACTORS MODIFYING FREQUENCIES

We have established the patterns which would be set up for autosomal and sex-linked loci if nothing happened to disturb the system. These patterns are ideal; we are more interested in the ways in which disturbance occurs. In the 1930s Sewall Wright produced an account of the factors changing gene frequency, referring to the change as the elementary evolutionary process. He classified factors into three groups, depending on what could be known about the change. These are:

1. the dispersive process, for which the size of the change is in principle determinate but its direction is indeterminate;
2. systematic processes, for which both size and direction are in principle determinate;
3. non-recurrent effects, for which neither size nor direction is determinate.

These will be considered in turn, and the criteria for classifying them in this way will be explained.

7.4 DISPERSIVE EFFECT. SIZE DETERMINATE IN PRINCIPLE, DIRECTION INDETERMINATE

The dispersive process is variously known as random drift, genetic drift or random sampling error. It is due to the fact that populations are not infinite, so that random fluctuation in gene frequency occurs, depending on the size of the sample which forms each new generation. Most organisms produce far more gametes than can survive if the population is to remain stable, so that formation of the new set of zygotes involves a sampling process.

Suppose we have an animal which produces a large number of gametes, and they carry alleles A and A′ at frequencies $p = q = 0.5$. The simplest way of illustrating the dispersive process would be to consider what would happen if the population was continued by just two surviving adults. Each would have two alleles, which would be AA, AA′ or A′A′ with probabilities p^2, $2pq$ and q^2. These probabilities are obtained by multiplying out the expression $(p + q)^2$. The probabilities are independent in the two individuals, so that the probability that the population is continued by a mating of two A′A′ individuals would be the product of their individual probabilities of occurrence or q^4. The full range of possibilities is found by multiplying out, or expanding, $(p + q)^4$. This provides

no. of A′ gametes	0	1	2	3	4
probability	p^4	$4p^3q$	$6p^2q^2$	$4pq^3$	q^4
when $q = 0.5$	1/16	4/16	6/16	4/16	1/16

If we consider a series of populations, each of which was continued by drawing a single pair of individuals, these would also be the proportions in which the different combinations occurred.

If the population size, N, was 5, so that the number of gametes, g, was 10, and $q = 0.5$ then the equivalent distribution is

no. of A′ gametes	0	1	2	3	4	5	6	7	8	9	10	
probability	(1	10	45	120	210	252	210	120	45	10	1	)/1024

For any value g these probabilities may be calculated, because the chance of getting a gametes in g is given by

$$f_a = \binom{g}{a} p^a q^{g-a}$$

$\binom{g}{a}$ is the number of permutations in which a gametes may be chosen from g and is calculated as $g!/a!(g-a)!$. ($g!$ is read as 'g factorial' and is the product $g(g-1)(g-2)(g-3)\ldots 1$).

Comparing these two distributions we can see that the larger g is, the wider is the possible range of values a may take. The fluctuation in gene frequency from generation to generation is inversely related to g. In practice, its size is expressed by finding the frequency of alleles making up the next generation, and its variance.

If we take $g = 10$ and $q = 0.5$, the expected mean number of A′ alleles is $qg = 5$. The possible deviations are

no. of A′ gametes	0	1	2	3	4	5	6	7	8	9	10
deviation $(x - qg)$	−5	−4	−3	−2	−1	0	1	2	3	4	5

The variance may then be calculated as $\Sigma f_i(x_i - qg)^2$, where f_i is the probability or frequency of occurrence. In this case the sum comes to 2.5. Calculation of the variance of a binomial distribution is made easier, however, by the fact that for all possible values of q and g, the variance is pqg, which is the same as for the calculation outlined above.

To sum up, the statement above tells us that if 10 gametes are picked from a large number of available gametes in which the frequency of type A′ is 0.5, then on average 5 of the 10 will be of type A′ with a standard deviation of $\sqrt{pqg}$, or $\sqrt{2.5}$. Similar statements could be made for any other sample size or frequency.

In order to examine change in gene frequency, we want to work in terms of the frequency q, rather than the number qg. The standard deviation of q is the standard deviation of qg divided by g, i.e.

$$\frac{1}{g}\sqrt{pqg} \text{ or } \sqrt{\frac{pq}{g}}$$

This is a measure of the fluctuation in frequency to be expected as a result of taking a sample of size g.

A diploid population of N individuals is formed from $2N$ gametes, so that if an allele has the frequency q in generation 0, the frequency in generation 1 will be $q \pm \sqrt{pq/2N}$. (Note that the quantity $2N$ turns up repeatedly in population genetics. For the sake of typographical tidiness the reciprocal of $2N$ is written $1/2N$ in the text.)

7.4.1 THE CONCEPT OF EFFECTIVE POPULATION SIZE

In relaxing the conditions for Hardy–Weinberg equilibrium all we have done so far is to accept that N is not infinite. There are numerous other implicit assumptions made when considering the effect of sampling error. In particular, it is assumed that the sexes are equally common, that N remains unchanged from generation to generation and that the number of offspring per family is a random variable. Deviation from these assumptions will lead to greater fluctuations in gene frequency than predicted by the binomial standard deviation. One could try to estimate how much larger the fluctuations would be. Sewall Wright chose the opposite course, and developed the concept of effective population size, N_e. This is the size of a population which would have the same sampling error as an observed population N, if all the conditions were properly obeyed. Almost always, N_e is smaller than N, sometimes a lot smaller. Expressions have been derived to estimate N_e.

Unequal sex ratio

If the sex ratio is unequal at mating, which is often the case in natural populations, the effective number is given as

$$N_e = \frac{4N_m N_f}{N_m + N_f}$$

where the subscripts refer to males and females, respectively. A population of 100 males and 100 females would have an effective size of 200. If there were 180 females and only 20 males, however, the effective size would be 72.

Change in *N* from generation to generation

When the population size fluctuates the effective population size is the harmonic mean of the numbers in each generation, i.e.

$$N_e = \left[\frac{1}{n}\left(\frac{1}{N_1} + \frac{1}{N_2} + \cdots \frac{1}{N_n}\right)\right]^{-1}$$

for n generations. The effect of this averaging is that small population sizes have a marked effect in depressing N_e. For the sequence 100, 100, 5, 100, 100 the arithmetic mean is 81, but the harmonic mean, and N_e, is only 21.

Variation in family size

If the numbers remain constant from generation to generation then on average each pair of parents should contribute two offspring to the next generation. Of course, there will be some variation about this mean, but the starting assumption is that the variation will have a Poisson distribution with a mean and variance of 2. If the variance is greater than 2 then N_e will be depressed. The effective population size is then obtained from

$$N_e = \frac{4N - 4}{V_2 + 2}$$

where $V_2 = s(1 - s)\bar{x} + s^2V$ and V = variance of family size, $\bar{x}$ = mean of family size, $s = 2/\bar{x}$.

It is usual for the variance of family size to be greater than the mean. The kind of effect this will have is illustrated in some data from a student project in which population number and family size were studied in the inhabitants of the Hebridean island of Eriskay. The number of people on this island increased from a handful in the mid-eighteenth century to over 500 in the early part of the twentieth century and thereafter dropped again. The number of inhabitants and the number of children per family both changed. In Table 7.2 the numbers are divided into 20-year periods.

When the population (N) was large, so was the family size and its variance also. Some families would have only one child while others had

Table 7.2 Variations in family size and population number of the Hebridean island Eriskay over three 20-year periods

Period	1891–1910	1911–1930	1931–1950
$\bar{x}$	5.0	4.6	3.0
V	11.8	8.3	3.0
N	466.0	540.0	380.0
N_e	367.0	381.0	380.0

many. The effect is to depress N_e. By the last period, however, the mean family size had gone down and there is less variation between families. N_e is more or less constant throughout, so that the size of sampling error is unaffected by the changes which have occurred.

7.4.2 INBREEDING

One of the dangers for small populations living on islands is that they become inbred, so that the frequency of deleterious recessive conditions manifesting themselves increases. Each individual in any population has two parents, four grandparents, eight great grandparents, etc. The number of direct ancestors doubles with each generation back, until it exceeds the population size N. At some stage all individuals have common ancestors. The smaller N is, the sooner a given degree of relatedness is reached, i.e. the more inbred the population is.

A consequence of this is that if we consider a series of similar finite populations, or a series of loci in a single population, or the probability distribution for a single locus in a single population, then the likelihood of observing homozygotes increases with time. We have seen this effect when there is complete inbreeding (selfing) in section 5.6. It also occurs on average with other breeding systems.

If an individual has two copies of the same allele (A′, say) then they may be identical because one has mutated to the A′ form from some other form or because the alleles come from the same ancestor and are identical by descent. The chance that a gamete will unite with an identical one is called the inbreeding coefficient or fixation index, and is represented by the letter F.

The size of F can be seen by considering an ideal population of hermaphrodites capable of self-fertilization, whose gametes mix at random. Imagine a starting situation where each individual produces gametes with one or other of two different alleles at a locus, and no individual carries alleles in common with any other. There are therefore multiple copies of $2N$ alleles present, from which the $2N$ gametes forming the next generation are chosen at random. Any gamete then has a probability of $1/2N$ of uniting with another of the same type. The population started, by definition, with $F = 0$, and the degree of inbreeding which has occurred is $1/2N$.

In the next generation gametes again have a probability of $1/2N$ of uniting with another of the same type (representing new inbreeding). A fraction $1 - 1/2N$ of different gametes is drawn in this round. However, it has already been shown that there is a probability of $1/2N$ that they are identical as a result of the previous round. The total probability of identity after two generations is therefore,

$$F_0 = 0$$

$$F_1 = \frac{1}{2N}$$

$$F_2 = \frac{1}{2N} + \left(1 - \frac{1}{2N}\right)\frac{1}{2N}$$

or

$$\frac{1}{2N} + \left(1 - \frac{1}{2N}\right)F_1$$

In general, for any generation n,

$$F_n = \frac{1}{2N} + \left(1 - \frac{1}{2N}\right)F_{n-1}$$

If we start with $F_0 = 0$, we can see that

$$F_1 = \frac{1}{2N}$$

$$F_2 = \frac{1}{2N} + \left(1 - \frac{1}{2N}\right)\frac{1}{2N} = 1 - \left(1 - \frac{1}{2N}\right)^2$$

$$F_3 = \frac{1}{2N} + \left(1 - \frac{1}{2N}\right)\left[\frac{1}{2N} + \left(1 - \frac{1}{2N}\right)\frac{1}{2N}\right]$$

$$= 1 - \left(1 - \frac{1}{2N}\right)^3$$

and

$$F_n = 1 - \left(1 - \frac{1}{2N}\right)^n$$

$$= 1 - e^{-n/2N}$$

After $2N$ generations $F_n = 1 - e^{-1} = 0.63$, i.e. for any population, 63% of complete inbreeding is achieved in $2N$ generations.

The inbreeding coefficient therefore increases progressively from zero, when no alleles are identical by descent, to 1, when they all are. The rate of inbreeding is inversely related to N. These are ideal conditions where self-fertilization can take place. More realistic conditions alter the rate, but only by a small amount, and the above equations can be taken to indicate the general effect of inbreeding.

In section 7.4 we saw that if a population of N individuals with two alleles at a locus is started with gene frequency q_0, the variance of gene frequency after one generation is $p_0q_0/2N$. In generation 1 the expected frequency will be expected to deviate from q_0, with a probability measured by this variance. In generation 2 the expected frequency will differ slightly from that in generation 1, and on average, rather more from the frequency in generation 0. Thus, if we measure deviation, or variance, from the fixed starting frequency, we see that it must increase cumulatively with generation.

The increase in variance is really another way of expressing the process resulting in increase in inbreeding, and the change occurs at the same rate. It can be shown that in terms of q_0 the variance in generation n is

$$V_n = p_0 q_0 F_n$$
$$= p_0 q_0 \left[1-\left(1-\frac{1}{2N}\right)^n\right]$$

Whatever the starting frequency, the probability that the gene frequency reaches 0 or 1 and becomes fixed at that point increases progressively. For $q_0 = 0.5$, there is an equal probability that the frequency will be anywhere in the range 0 to 1 after $2N$ generations, and thereafter the probability of fixation is $1/2N$ per generation. If q_0 has an extreme value (near 1 or 0) it takes nearer $4N$ generations for this condition to be achieved. The probability distribution of q when the number of generations is measured in units of N is shown in Figure 7.2.

Within a generation, the disturbance of the average Hardy–Weinberg expectation for two alleles can be represented in terms of F as shown in Table 7.3.

In this table the Fs refer to the value F has attained in the particular

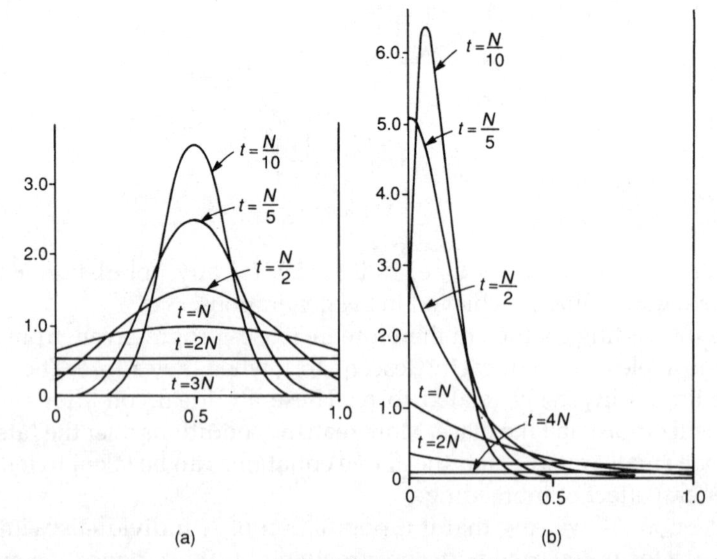

Fig. 7.2 The process of random genetic drift due to a small population number, in which it is assumed that mutation, migration and selection are absent, and the random change in the gene frequency from generation to generation is caused by random sampling of gametes. In (a) the initial frequency of q is 0.5, while in (b) the initial frequency is 0.1. In (a) and (b), t stands for the time and N stands for the effective population number. The ordinate is the probability of having a value q. From Kimura (1955).

generation reached. As time proceeds, F becomes larger and the probability that the population is homozygous becomes greater. We will return to this point when discussing random models for polymorphism.

7.5 SYSTEMATIC EFFECTS. SIZE AND DIRECTION BOTH DETERMINATE IN PRINCIPLE

The second major category on the Sewall Wright classification includes three distinct effects: migration, mutation and selection. All of them may change gene frequencies by amounts which can in theory be measured, and unlike sampling error, it is also possible to say in which direction the frequency will be pushed.

7.5.1 MIGRATION

Many detailed studies of migration in real animals and plants have been made, often with quite complicated results. Using the model of separate demes with a small number of migrants between them, the general principle may be stated quite simply. For any one population, immigrants arrive at a rate m per generation, which is the number of immigrants divided by the total number in that deme after the immigration has occurred. For a given locus, the frequency of one allele in the population (q, say) may be different from that found in the immigrants (q_m, say). We are interested in how immigration changes gene frequency. After one generation has elapsed the new frequency is

$$q_1 = (1 - m)q_0 + mq_{m0}$$

that is to say, the sum of the frequency among non-immigrants and the frequency among immigrants.

Since we are interested in change in gene frequency, it is convenient to express the change as

$$\begin{aligned} \Delta q &= q_1 - q_0 \\ &= q_0 - mq_0 + mq_{m0} - q_0 \\ &= m(q_m - q), \text{ leaving out the subscripts} \end{aligned}$$

This is the simplest representation of the effect of migration, stating that the change in frequency brought about is the product of migration rate and the difference in frequency between immigrants and indigenous individuals.

Table 7.3 Effect of inbreeding on expected genotype frequencies in a population of finite size

Genotype	Hardy-Weinberg expectation	Effect of inbreeding
AA	p^2	$p^2 + pqF$
AA′	$2pq$	$2pq - 2pqF$
A′A′	q^2	$q^2 + pqF$

7.5.2 MUTATION

The variability which makes evolution possible arises through mutation. Some mutations may be so rare that they are, in effect, unique – in which case nothing can be said about their dynamics. Others are common enough to be said to have a rate of occurrence. The rate is small; figures of around 10^{-9} per locus per generation occur in bacteria and as much as 10^{-4} for some loci in man. The differences are in part accounted for by difference in generation time and in part by differences in definition.

Any chromosomal change may be said to be a mutation, including, for example, large deletions or other chromosomal rearrangements. At the other extreme, a single base within a DNA sequence may be converted to another base. At the molecular level, an average gene may be 1000 base pairs long. Any one of these may mutate to one of three other forms. There are four different forms possible for each site, and therefore 4^{1000} or 10^{602} different combinations of bases in the gene. At the DNA level, the number of gene mutations is effectively infinite.

At the phenotypic level, however, it is quite common to have a situation where a typical form prevails but a small number of mutant types is known. The golden hamster, well known as a pet and laboratory animal, was domesticated from a single family brought from Syria in the 1930s. At first, all hamsters had the same coat colour, but as time passed and numbers in captivity increased mutant forms came to be observed and saved until now about a dozen are present.

Sometimes revertants from mutant to wild type occur, although these are not commonly detectable. One possible example is seen in the ferret, which is a domesticated albino version of a wild polecat which has a cream and brown patterned coat. So-called polecat ferrets are probably revertants to the original wild type.

Because of the difference in our understanding of the phenotypic and molecular level, there are two types of model for the dynamics of mutation, which may be called the 'back mutation' model and the 'infinite allele' model. The first was discussed by Sewall Wright while the second was due to the Japanese geneticist Motoo Kimura. In the infinite allele model, the fate of each individual mutant is unpredictable but the net effect on heterozygosity can, in principle, be determined.

Back mutation model

Suppose a gene A mutates to a form A′ at rate v per generation, and the frequency of A is q, while that of A′ is p. Then

$$q_1 = q_0 - vq_0$$

and

$$\Delta q = -vq$$

However, back mutation may occur, at rate u, so that the complete process is represented by

$$q_1 = q_0 - vq_0 + up_0$$

and

$$\Delta q = up - vq$$

The two forces represented by up and vq must both be zero or positive; they can never be negative. Since they act in opposite directions, it is possible for them to balance each other at some frequency so as to produce an equilibrium. At equilibrium $\Delta q = 0$, that is to say, no change in gene frequency occurs. When $\Delta q = 0$

$$up - vq = 0$$

therefore

$$u - uq - vq = 0$$

so that

$$\hat{q} = \frac{u}{u+v}$$

and $\hat{q}$ is the equilibrium frequency. The relation of Δq to q is shown graphically in Figure 7.3.

As a result, mutation keeps both alleles in the population, at a frequency which depends on the mutation and back mutation rates. However, the frequency moves towards a frequency determined by the higher rate. Since this is almost certainly the mutation rate rather than the back mutation rate (since misreplication is more likely than repair), the population should end up with more of the so-called mutants than of wild types from which they mutate. It does not, otherwise we should not call the alleles wild type and mutant. Something else, presumably selection, must eliminate the mutants.

Infinite allele model

A gene may consist of so many combinations of bases that many, if not most, must have essentially the same effect. On this assumption, most mutants are neutral in effect. A mutant is a change, usually at a single base, in a cell lineage. If such a mutation arises, at frequency $q_0 = 1/2N$, it will become extinct or increase in frequency by chance if it is neutral. If the mutant increases in frequency until it becomes fixed, then there is said to be a substitution. The smaller the population the greater the chance that it will be fixed, and mathematical analysis shows that the chance of fixation is q_0 or $1/2N$.

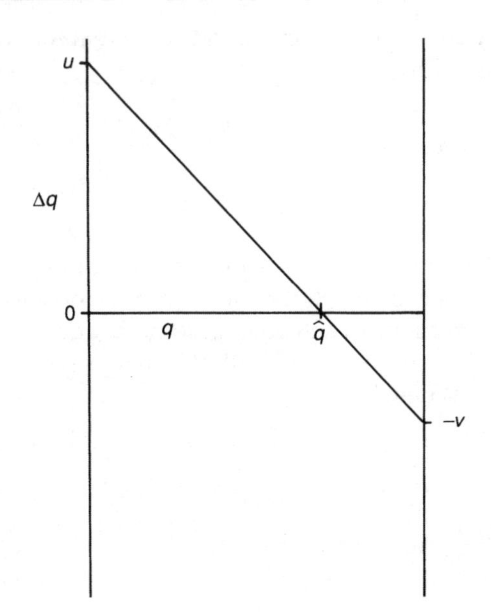

Fig. 7.3 The relation of Δq to q for mutation and back mutation. From the equation $\Delta q = u$ when $q = 0$ and $-v$ when $q = 1$. The slope of the line is $-(u+v)$ and at equilibrium $\hat{q} = u/(u+v)$. The figure shows that any perturbation above q will be followed by a decline in q (Δq −ve) and one below q will be followed by an increase (Δq +ve). The equilibrium is therefore stable.

Proper demonstration of this involves some more advanced mathematics than is used here, but the following argument shows that it should be so. In section 7.4 we saw that if we start with a frequency q_0, the variance increases with generation as a result of drift, until the allele is either lost from the population or fixed. If we start with $q_0 = 0.5$, as in the example given, then these alternative outcomes are equally likely. If q_0 is less than 0.5 then loss is more likely than fixation, and the probability of the alternative end points is inversely proportional to the starting frequency. A starting frequency of 0.25 means that the probability of loss is 0.75 and that of fixation 0.25, a starting frequency of 0.1 ensures that the chance of fixation is 0.1. By the same argument, the probability of fixation of an allele which enters the population at a frequency of $1/2N$ must itself be $1/2N$.

In any generation the number of new mutants arising is $2Nu$, where u is the mutation rate per generation. The probability that a mutant arises in a given generation and subsequently becomes fixed is, therefore, the product of these two quantities, i.e. $2Nu \times 1/2N = u$. The conclusion is that gene substitution occurs at a constant rate, equal to the mutation rate u, which is independent of population size. On average substitutions occur once per $1/u$ generations. This conclusion is based on the assumption that the mutants are neutral.

For the following reasons, the continual production of new mutants can lead to an equilibrium state in which there are always several present in a population. The transition from mutation to fixation is a change from a frequency of $1/2N$ to $(2N - 1)/2N$. If a population is large, this will take longer than if it is small, and as one mutant is travelling this course, others may arise.

The time required to move from mutation to fixation can be shown to be $4N$ generations on average. The probability of a substitution, however, is u per generation, so that substitutions will occur, on average, once per $1/u$ generations. If $4N$ is small compared with $1/u$ the population will be invariant (monomorphic), while if it is large the population will contain many variant alleles. We can rewrite this by saying that the population will be monomorphic if $4Nu \ll 1$ and polymorphic if $4Nu > 1$. A powerful body of theory can be established on this basis, which allows prediction of patterns of distribution of gene frequencies. We shall return to the Kimura infinite allele model in Chapter 12.

7.5.3 SELECTION

The sequence of events occurring during the life cycle is shown in diagrammatic form in section 4.9. If adjustment of numbers occurs differently in different genotypes at any of the stages shown, then selection has occurred. Sometimes we say there has been selection simply because the change has been observed to take place, but very often it is possible to identify a cause of selection and study the change it produces.

Gametic selection

In order to see how selection affects relative numbers, it is easiest to start with consideration of gamete production. Suppose there are two types of gamete carrying allele A or A′. Out of every 100 A gametes 90 survive to the time when they can form zygotes. Out of 100 A′ gametes, 80 survive. In order to represent the change in terms of gene frequency, so that a frequency q_0 changes to q_1, we would have

$$\frac{100}{100+100} \rightarrow \frac{80}{80+90}$$

Description of selection in terms of gene frequency always involves a quotient, making it algebraically more complex than the processes considered so far.

Selection is measured as fitness, or selective value, represented in equations by w. In this case

$$100\, w_1 = 90$$
$$100\, w_2 = 80$$

For g gametes, this would be

$$pgw_1 = 90$$
$$qgw_2 = 80$$

where $p = q = 0.5$ and $g = 200$, and consequently we can write

$$q_1 = \frac{q_0 w_2}{p_0 w_1 + q_0 w_2}$$

since the gs drop out of the equation. This method of representing selection loses all information about g (or N); the numbers involved could be increasing, decreasing or remaining constant. The selective values therefore represent fitness of one type relative to the other, and the equation can be simplified without loss of information by setting the first fitness value at unity, so that the second will be w_2/w_1. The equation becomes

$$q_1 = \frac{wq_0}{p_0 + wq_0}$$

To measure change in gene frequency, we then have

$$\Delta q = \frac{wq}{p + wq} - q$$
$$= \frac{-pq(1-w)}{p + wq}$$

The denominator in such an expression is called the mean fitness, denoted by $\overline{w}$, since it is the product of the frequencies of the different genotypes and their relative fitnesses. Another piece of terminology frequently used is to define the selective coefficient s as $1 - w$. The selective coefficient has a direct meaning; $s = 0.2$ means that there is a 20% disadvantage to the type concerned compared with one of the other types in the population. Using these two conventions, the equation becomes

$$\Delta q = \frac{-s\,pq}{1 - sq}$$

Can there be an equilibrium with gametic selection? We answer this by finding the conditions under which $\Delta q = 0$. This occurs, when the numerator equals zero, i.e. when $s = 0$, so that there is no selection, or when p or q equal zero. Equilibrium therefore requires that there is no choice, or when there is a choice, no selection. Selection will therefore move the gene frequency in one direction or the other. The trajectory of change is S-shaped. It is slow when q is near zero or unity and more rapid near the middle of the gene frequency range. This pattern is illustrated in Figure 7.4 by the graph of Δq on q.

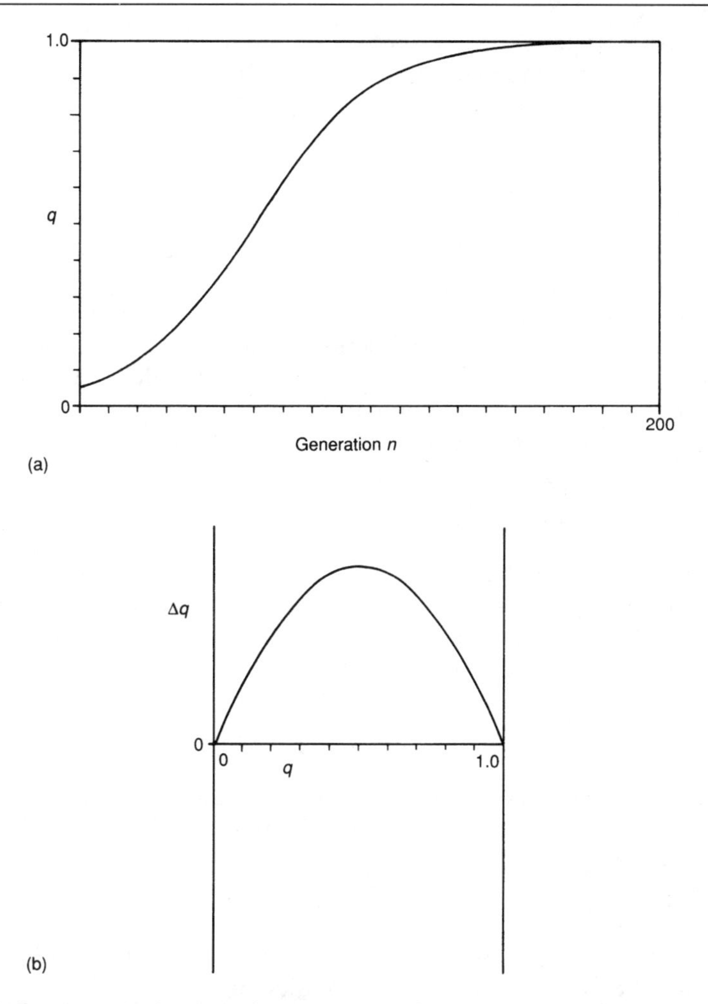

Fig. 7.4 Gametic selection favouring a gamete at frequency q. (a) shows the S-shaped curve of frequency on time. (b) shows change in frequency (Δq) plotted on frequency. There can be no selection when there is no choice, so $\Delta q = 0$ when $q = 0$ or $q = 1$. Between these values, the gene always increases in frequency in this case, since it is advantageous. The maximum rate of change is at $d\Delta q/dq = 0$. With 5% advantage the scale shows 200 generations.

Zygotic selection

The one-locus two-allele system in a large population starts with formation of zygotes more or less in Hardy–Weinberg proportion. After that, selection may lead to differential mortality of genotypes. In a manner analogous to that for gametic selection, the situation may be represented as follows.

Genotype	AA	AA′	A′A′
Starting frequency	p^2	$2pq$	q^2
Relative fitnesses	w_1	1	w_3
Final frequency	$\frac{w_1 p^2}{w}$	$\frac{2pq}{w}$	$\frac{w_3 q^2}{w}$

The fitness of the heterozygote, w_2, has been set at 1. The mean fitness $\overline{w}$ is $w_1p^2 + 2pq + w_3q^2$. The ws in the above notation are related to the genotypes whereas in the foregoing discussion we have talked about factors affecting gene frequency. To do the same with zygotic selection, we would need factors x_1 and x_2 which measure fitness of the alleles at frequencies p and q. Values for x_1 and x_2 could be found as follows. The alleles at frequency p are in the population in the combinations w_1p^2 and pq, while the others are in the combinations w_3q^2 and pq, so

$$w_1p^2 + pq = p(w_1p + q) = p(1 - s_1p) = x_1p, \text{ say, and}$$
$$w_3q^2 + pq = q(w_3q + p) = q(1 - s_3q) = x_2q$$

Thus, x_1 and x_2 are both functions of gene frequency. Even though fitness values of the genotypes are constant, selection on the genes varies in strength with frequency. To obtain the frequency in the next generation, we write

$$q_1 = \frac{x_2 q_0}{x_1 p_0 + x_2 q_0}$$

which is the same form as for gametic selection except that x_1 and x_2 themselves vary with frequency. The change in frequency is

$$\Delta q = \frac{-pq(x_1 - x_2)}{x_1 p + x_2 q}$$

which again is the same in form as for gametic selection. Substituting $1 - s_1p$ for x_1 and $1 - s_3q$ for x_2, the expression becomes

$$\Delta q = \frac{pq(s_1 p - s_3 q)}{1 - s_1 p^2 - s_3 q^2}$$

The trajectories of gene frequency change with time are S-shaped, as they were for gametic selection. Selection in favour of a dominant gene may be represented by setting s_3 at zero and making s_1 positive. It starts off rapidly at low frequencies, then slows down as more and more of the remaining unfavoured alleles are present as heterozygotes, so that the final move to fixation takes a long time. On the other hand, selection in favour of a recessive ($s_1 = 0$ and s_3 negative) produces a very slow response at low frequencies, but once the majority of alleles are of the favoured type there is a rapid move to fixation. These different patterns of response are shown in Figure 7.5.

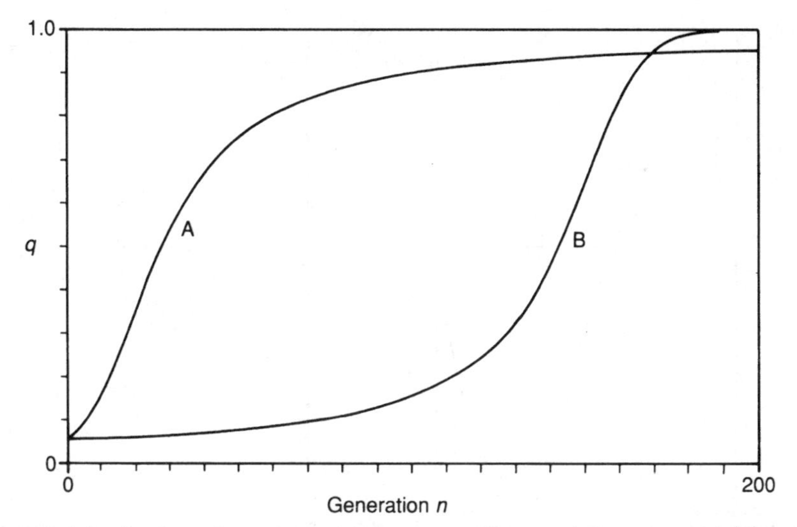

Fig. 7.5 Selection favouring a gene at frequency q. Characteristic patterns of change in frequency when the gene is (A) dominant and (B) recessive. In the example shown, the selective advantage is 0.15 and $q_0 = 0.05$. In each case the trajectory is sigmoid. Response of a dominant starts rapidly but completion of the change takes a long time. Response of a recessive is very slow at first, but when q is large enough for there to be an appreciable frequency of homozygotes in the population the gene rapidly goes to fixation.

As before, equilibrium will occur if p is zero, q is zero or the part in brackets is zero. The last condition will hold if $s_1 p = s_3 q$,

$$\therefore\ s_1 - s_1 q = s_3 q$$
$$\therefore\ \hat{q} = \frac{s_1}{s_1 + s_3}$$

The condition for equilibrium looks the same as that for forward and back mutation in section 7.5.2. It is not identical, however, because s_1 and s_3 may each be positive or negative, whereas u and v may only be positive if not zero. If one selective coefficient is positive and the other negative then there is no equilibrium and the frequency moves to zero or unity. If they both have the same sign an equilibrium is possible, but the conditions of stability differ. This may be seen by looking at the graphs of Δq on q (Figure 7.6).

In Figure 7.6(a) s_1 and s_3 are both positive and the curve crosses the $\Delta q = 0$ line with negative slope at $\hat{q}$. A perturbation will be followed by a return to the equilibrium frequency, and the equilibrium is therefore stable. In Figure 7.6(b) s_1 and s_3 are both negative, the slope is positive and a perturbation is followed by ever-increasing divergence until 0 or 1 is reached. The equilibrium is unstable. The first of these situations is heterozygote advantage and the second is heterozygote disadvantage. A similar pair of outcomes was seen in the competition equations in section 4.3.

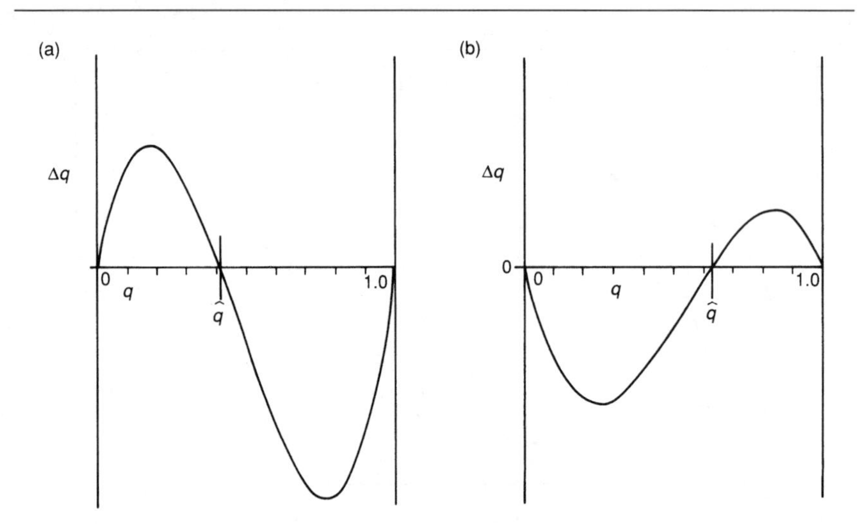

Fig. 7.6 Equilibrium with zygotic selection and constant fitnesses. There is no change in gene frequency if $q = 0$, $q = 1$ or $q = \hat{q}$. Graph (a) represents heterozygote advantage. If the frequency is perturbed from $\hat{q}$, it will move back towards it. Graph (b) represents heterozygote disadvantage. After perturbation from $\hat{q}$, the frequency will move towards $q = 0$ or $q = 1$ depending on whether the frequency is below or above $\hat{q}$.

Selection favours the fittest types, and therefore tends to increase their frequency. As a result it tends to increase the mean fitness, measured by $\overline{w}$ in the Δq equations. In Figures 7.4 and 7.6, selection with constant fitness values for the three genotypes formed by two alleles at a locus is illustrated as the relation of Δq to q. Figure 7.7 shows the same process represented as the relation of $\overline{w}$ to q. Actually, study of more complex selection models (which may have frequency-dependent genotype fitnesses or with locus interaction) shows that $\overline{w}$ does not always maximize. Nevertheless, the principle is sufficiently general for it to be possible to visualize the result of many types of selection as increase in $\overline{w}$.

Examining the figures, we can see that in Figure 7.7(a) the frequency will move towards $q = 1$, where the allele with the highest selective value is fixed. In Figure 7.7(b), the frequency moves to $q = \hat{q}$, the stable equilibrium point. It is possible to rewrite the Δq equations so that they tell us that $\Delta q = 0$, the equilibrium state, when $q = 0$ or $q = 1$ or $\mathrm{d}\overline{w}/\mathrm{d}q = 0$, and the mode in Figure 7.7(b) represents the latter type of equilibrium point. In Figure 7.7(c), the frequency moves to $q = 0$ or $q = 1$, depending on whether the starting value of q is above or below the unstable equilibrium point, $\hat{q}$. We therefore have the possibility of two different degrees of adaptation, one of which is higher than the other. The one which is actually reached depends on the point from which the population starts.

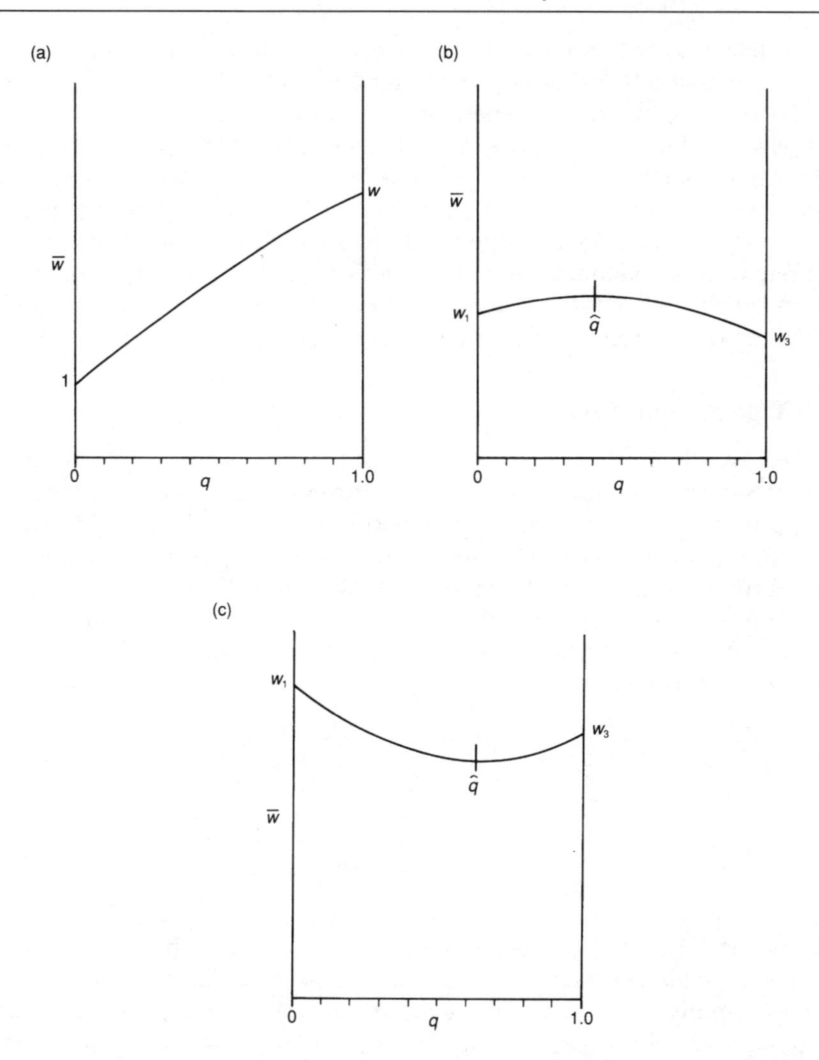

Fig. 7.7 Relation of mean fitness $\hat{w}$ to q. This is another way of representing the information in Figures 7.4 and 7.6, which emphasizes the effect selection has on average fitness. (a) Gametic selection favouring the allele at frequency q; (b) zygotic selection, heterozygote advantage; (c) zygotic selection, heterozygote disadvantage. In (b) and (c), $\hat{q}$ is the value for which $d\hat{w}/dq = 0$.

Selection on quantitative characters

It is assumed that continuously varying characters are controlled by segregating genes at several loci, sometimes, but not always, concerned primarily with the character studied. It is not possible to unravel the details of the control, however, and assign fitness values to the genotypes at each locus. For these characters, it is necessary to rely on the phenotype selection methods discussed in Chapter 4, section 4.7.

Sometimes, theoretical geneticists have developed multilocus models of selection, but the method is of limited value for obtaining explicit solutions to questions of selection patterns, equilibria, etc. For two loci each with two alleles, there are nine genotypes (or 10, if the coupling and repulsion heterozygotes are distinguished), each of which could be assigned a different fitness value. With more alleles or loci, the number of combinations quickly becomes too large to provide meaningful insights, and we shall restrict ourselves for the most part to one locus.

Linkage disequilibrium

When two loci are studied, one more quantity is frequently used to express the interaction. This is D, the coefficient of linkage disequilibrium. Suppose the two loci have alleles A and A′ at frequencies p and q, and B and B′ at frequencies r and s. If there is no disequilibrium, the frequency of the chromosome combinations AB, A′B, AB′ and A′B′ will be the products pr, qr, ps and qs. With disequilibrium, some are in excess and some deficient compared with the expectation, so that the four combinations are represented as follows.

chromosome type	frequency
AB	$a = pr + D$
A′B	$b = qr - D$
AB′	$c = ps - D$
A′B′	$d = qs + D$

If the frequencies a, b, c and d are known then D may be calculated, since $D = ad - bc$. If $D = 0$ then there is no disequilibrium, but when D is positive or negative the genotype of an individual in a population cannot be predicted simply by knowing the gene frequencies at the individual loci. Non-zero D values may indicate that a population was started by a few individuals in linkage disequilibrium and that equilibrium has not yet been reached or that selection favours particular combinations of alleles at the two loci at the expense of others. Selection which maintains disequilibrium is epistatic in its effect. We usually see evidence of disequilibrium when loci are closely linked, but in fact no physical linkage is necessary for selection to generate a non-zero value of D. For that reason gametic excess would be a better description, but the term linkage disequilibrium seems to have stuck.

The rate of reduction of D in the absence of selection depends on linkage distance. From one generation to the next, $D_{n+1} = D_n(1 - R)$, where R is the recombination frequency (cross-over value), or $D_n = D_0(1 - R)^n$. The number of generations required for D to halve can be found by writing

$D_n/D_0 = (1 - R)^n = \frac{1}{2}$. Converting to logarithms we get $n\log(1 - R) = \log\frac{1}{2}$ or $n = \log\frac{1}{2}/\log(1 - R)$. This is 1 generation if $R = 0.5$ (that is, when there is 50% crossing-over), but 700 generations when $R = 0.01$. If there is close linkage, a population started by a few individuals with extreme disequilibrium could remain in that state for hundreds of generations, so that knowledge of the linkage distance is essential before anything may be said about the possible involvement of epistatic selection.

The 10 genotypes arising from two alleles at each of two loci have the following frequencies.

		locus A		
		AA	AA′	A′A′
	BB	a^2	$2ab$	b^2
locus B	BB′	$2ac$	$2ad + 2bc$	$2bd$
	B′B′	c^2	$2cd$	d^2

The frequencies in the central cell represent coupling and repulsion heterozygotes, A/A′ B/B′ and A/A′ B′/B.

In principle, each of these 10 genotypes could have a different selective value so that complicated patterns of selection may be generated, and various models have been investigated theoretically. Extreme values of D are likely to indicate that diagonal pairs of homozygous combinations (e.g. those represented by a^2 and d^2) are favoured over the alternatives (b^2 and c^2). In human genetics, linkage disequilibrium is seen in the MN and Ss blood groups and between some of the loci in the HLA histo-incompatibility system.

If linkage disequilibrium is generated by selection, then automatically there is also selection for closer linkage. This is because an individual in which the less favoured genotypes do not segregate is fitter than one in which they do. Any mechanism which reduced the amount of recombination, such as non-reciprocal crossing-over, translocation or inversion of a segment of chromosome, would be favoured. Such selection would operate against deleterious effects which may accompany the chromosomal change. It is also much less likely to be successful if the loci are loosely linked than if they are tightly linked. Chromosomal reorganization driven by epistatic selection is therefore likely to be rare, but in the long term, this process could have a significant effect. Examples in which such selection may play a part are found in the inversion polymorphisms of *Drosophila* species, and one concerning mimicry in butterflies is considered in Chapter 9.

r and K selection

Change in gene frequency as a result of random drift will result in the phenotype changing with time, and in divergence between populations and species. It is not adaptive, however, and is different from the evolutionary process controlled by natural selection. Is selection 'Darwinian'? In that process the role of density dependence is emphasized, whereas selection as discussed so far in this chapter has been restricted to gene frequency change. Often the density-dependent component is unimportant and may safely be ignored, but the real reason for leaving it out has been that it greatly complicates the story, introducing parameters which are hard to quantify and making it difficult to derive testable predictions. This may have led to neglect by geneticists of important aspects of population dynamics.

The notation of the logistic growth curve has given rise to the terms r- and K-strategy to describe two extreme ways in which organisms may be adapted. The distinction was first made in this way by R.H. MacArthur. The r-strategist is a species with a very high rate of increase and high mobility, able to expand rapidly into new territory. It can occupy temporary habitats and reproduce sufficiently fast to continue to find new ones. Effective dispersal is important but competitive ability is not. A K-strategist, on the other hand, has to compete in a stable habitat, where it is established at a density near the saturation value K. Such a species will possess adaptations to increase its success in crowded conditions. It may be relatively large sized, and consequently long lived, and is likely to possess specific defences against competitors. The reproductive capacity will be lower than in an r-selected species, but there is more likely to be parental care. Birth rate is likely to be sensitive to density, so that increased reproductive output can compensate for a drop in numbers. Nevertheless, since the accent is on competitiveness, rather than output, a drastic reduction in numbers is likely to lead to extinction.

These generalizations help to describe the pattern of ecological characteristics which species possess. In plants, annual weeds are typical r-strategists, with small wind-borne seeds able to colonize newly available open ground. They grow and reproduce rapidly, but as the vegetation becomes established, are replaced by more efficient competitors. At the other extreme, tropical rain forest is composed of large long-lived trees which compete with each other to reach the light. They often produce large seeds with a high energy reserve which can grow very rapidly to fill a space left by a fallen old tree. They may be poor dispersers but possess effective defence mechanisms against insects and mechanisms which inhibit growth of neighbouring trees. Many geometrid moths living in forests are r-strategists. The adults do not feed and are little more than machines for producing large numbers of eggs. Newly hatched larvae disperse by producing silk thread parachutes on which they are blown about by the wind. Among those which land in suitable places, there is rapid growth, and all the energy required for future egg production is obtained through larval feeding. The other extreme is shown by heliconiid butterflies. These tropical forest dwellers live for a month or more as adults. They learn the distribution of nectar-producing flowers and of larval food vines. In many species, eggs are produced one at a time and laid on carefully selected tendrils or young leaves. The larvae pick up alkaloids from their food plants which make the adults poisonous, and the adults are warningly coloured to deter predators.

Of course, many species do not fit neatly into r or K categories and care has to be employed in deciding on the significance of a particular feature. In a region of oceanic islands, small wind-borne seeds would be a natural adaptation to expect in an r-strategist. Large seaworthy fruits such as coconuts might be equally effective dispersal agents, however, whereas in a forest context they would appear to indicate a K-strategy. One reason why the classification works at all is that organisms are subject to energy constraints. If there are large seeds there will be relatively few of them. If much energy is spent on defence mechanisms then correspondingly less is available for reproduction. There must therefore be some degree of negative association between the r features and the K features.

Within a species in which there is genetic variation, the fitness of each genotype may be thought of in r and K terms. Using the notation in Chapter 4, section 4.3, a haploid population in which there were two genotypes at frequencies p and q may be represented as

$$p_{n+1}N_{n+1} = p_n N_n R_1 \left\{1 - \frac{(p_n N_n + a_2 q_n N_n)}{K_1}\right\}$$

$$q_{n+1}N_{n+1} = q_n N_n R_2 \left\{1 - \frac{(q_n N_n + a_1 p_n N_n)}{K_2}\right\}$$

$$K_i = K'f(R_i)$$

The K value for each genotype is negatively related to its R value; K' is an overall value about which they are distributed. If a_1 and a_2 both equal 1 and there is a monotonic relation between K and R then the genotype with the highest K will increase in frequency and become fixed in the population. It has the greater competitive ability. When the R values are approximately the same, polymorphism could occur if $a_1 < K_2/K_1$ and $a_1 < K_1/K_2$. Additional individuals of each genotype would then have a greater inhibiting effect on further increase of their own than of the other genotype. This is likely to imply that they occupy at least partially separated niches. Alternatively, the relation of R to K could be such that K was always higher for the genotype at the lower frequency. There would then be frequency-dependent competitive ability. As in the two-species example, coexistence is more likely if there is some random mortality as well. In diploid organisms, heterozygous genotypes may have higher fitness than homozygous genotypes, which would also result in polymorphism.

7.6 NON-RECURRENT EVENTS. SIZE AND DIRECTION INDETERMINATE

This is the last of Sewall Wright's categories of factors affecting gene frequency. It contains anything which is so rare or improbable that it is impossible to describe it as having a rate. It includes non-recurrent, essentially unique, migration, mutation and selective events. It also includes the phenomenon known as the Founder Effect, which is a non-recurrent extreme reduction in numbers, such as might happen when a rare migrant colonizes an empty island and becomes the progenitor of a new and flourishing population. Such population bottle-necks would be accompanied by a radical change in genetic constitution, and their relative importance in evolution has frequently been discussed.

7.7 CONCLUSION

The purpose of this chapter has been to supply a set of tools, as it were, with which to examine patterns in real populations. The tool box is necessarily incomplete, but it should be possible to start to ask what kind of forces may be operating in a given case and to think of ways of recognizing the action of one kind of force, as opposed to another. The Hardy–Weinberg theorem gives us the basic expectation. If it is not met, one or more of the processes described must be operating. Of these, selection may be by far the most powerful and the most likely to be detected in the short term. Selection need not be strong, however, and mutation and drift may have small but inexorable effects on variability, mutation to generate it and drift to reduce it.

7.8 SUMMARY

Study of gene frequency change in populations usually starts from the Hardy–Weinberg principle that if nothing acts to modify the frequency at an autosomal locus it will remain constant with constant genotype frequencies which are the products of the frequencies among gametes. This equilibrium is stable and is reached after one generation. The factors which may change frequency are non-random mating and other selective processes favouring one allele over another, random drift, which is a consequence of the finite size of the population, migration and mutation.

Sewall Wright divided these factors into three groups, the dispersive process (drift) for which the size but not the direction of the change is in principle determinable, systematic processes (migration, mutation and selection), for which both size and direction of the change are in principle determinable, and non-recurrent events (non-recurrent selection, migration, mutation or reduction in numbers), for which neither can be measured. These distinctions are useful in analysing the consequences of the processes and in weighing up their relative importance.

Random drift is responsible not only for fluctuation in gene frequency, but at the same time, for increase in the probability that a locus will be homozygous (the inbreeding effect). Mutation may be described in terms of a few distinguishable phenotypes, in which case the dynamics operate in terms of rates of mutation to and from a given phenotype. In principle, it may also be considered at the molecular level, in which case a mutation is the replacement of one base by another, a gene may occur in an almost infinite number of base permutations and the probability of a back mutation is negligible. In this model, the predicted conditions are the probability that a given mutation will become fixed and the number of mutant types in a population at a given time.

Selection may take place at any stage in the life cycle, and may favour particular alleles or genotypes at the expense of others. Most selection is probably frequency dependent in its effect, or density dependent or both. The usual methods of describing selection make it difficult to recognize and to measure density-dependent selection, for which reason it is probably underrepresented in the literature. Heterozygote advantage and many types of frequency- and density-dependent selection may result in stable equilibria in gene frequency.

The standard algebra of gene frequency change is effective for describing processes at single loci. The theory of quantitative genetics provides an adequate account of the phenotypic effect of the combined action of many loci. The effect of selection, mutation and drift on a few interacting loci is, however, very difficult to describe adequately. This is unfortunate, since groups of several interacting loci are probably important in evolution.

7.9 FURTHER READING

Crow, J.F. (1986) *Basic Concepts in Population, Quantitative and Evolutionary Genetics.*
Crow, J.F. and Kimura, M. (1970) *An Introduction to Population Genetics Theory.*
Falconer, D.S. and Mackay, T.F.C. (1996) *Introduction to Quantitative Genetics.*

Genetic variability in natural populations 8

8.1 INTRODUCTION

The Darwinian theory of evolution by natural selection tells us that within the limits set by a changing environment, adaptation should always tend to increase. If the fittest increase it is because they are picked out of an array of types, including unfit ones which fail to survive. In the biological context, as in everyday usage, selection implies that a few are picked from a larger assemblage. Variability is therefore likely to decline. At its elementary genetic level, this implies the favouring of a particular allele over others at a given locus. If the selection is successful, a locus should remain monomorphic for the favoured allele until the environment changes in such a way that a different form is favoured. From its previous status as a mutant eliminated as it arose, this form then increases in frequency to become the new 'typical' gene at the locus in question. Thus, most loci should be monomorphic most of the time. If they are not, it is probably because replacement is taking place, and the most common number of alleles at such a locus should be two.

The central problem of population genetics is that this prediction is not borne out. A large fraction of the genome consists of loci at which at least two and often several alleles coexist. Of course, this variability could be useful in permitting rapid adaptation to new conditions, but there is nothing in the theory of evolution by natural selection, as it has so far been described, which says that it should be available for the purpose. R.C. Lewontin highlighted the problem by referring to the *paradox* of variation. We shall first look at the evidence and then at ways in which the paradox may be resolved.

8.2 VISUAL AND OTHER POLYMORPHISMS

Since the eighteenth and nineteenth century, there has been a tradition of natural history which has flourished in many countries, but perhaps especially in Britain, which has led to the recording of natural variation in animals and plants. We are accustomed to the fact that many animals have more or less constant appearance; that is true of common inhabitants of houses and gardens, such as many species of flies, wasps, beetles, slugs, spiders, etc. The *Drosophila* used in genetical experiments have a standard wild-type appearance. The genetics depends on rare mutants which have been noticed and saved in cultured stocks. Other familiar creatures such as dogs, cats and cattle are variable in appearance, but the variation is evidently the result of human breeding programmes.

Closer examination shows that a substantial fraction of wild animals and plants too have different variants present in populations. This is true of many butterflies and grasshoppers, and of moths, where dozens of species show melanic variants. Thistles vary in flower colour and clover in leaf patterns. Seashore molluscs, such as winkles, dog whelks and clams, coexist in different shell colour forms, and among land snails some groups have shell colour variation in 20–30% of species. Even when the morphology is invariant, as in the case of *Drosophila*, there may be hidden variation. For example, different alleles of the *Adh* gene code for different versions of the enzyme alcohol dehydrogenase, and can be distinguished either by the electrophoretic mobility of the proteins or by direct sequencing of the DNA. In general, 'fast' alleles which produce enzymes of high mobility, are associated with greater capacity to oxidize alcohols than 'slow' alleles. Most populations exhibit a diversity of *Adh* genotypes. Clover and vetches vary in their ability to produce cyanide when the leaves are broken, populations of wild thyme (*Thymus vulgaris*) vary in the secondary compounds which cause their fragrance. Sometimes a selective agent may be inferred, for example in the case of mimetic variants in butterflies, which are associated with selective predation, while in others the reason is completely unknown. Often the genetic basis has not been investigated, but this type of variation usually involves the coexistence of distinct phenotypes, and when they have been studied they are found to be controlled by segregating genes or groups of genes.

The presence of variants in a population is termed polymorphism. Thinking of examples such as those outlined above, F.B. Ford defined genetic polymorphism as 'the occurrence together in the same locality of two or more discontinuous forms of a species in such proportions that the rarest cannot be maintained by recurrent mutation'. This definition concentrates attention on the distinct phenotypes, rather than on the alleles which control them, and it excludes mutation as a means by which they develop. We have already seen that mutation could lead to a polymorphism (section 7.5.2) provided selection did not swamp its effect, so that the definition implicitly assumes the action of selection.

Naturalists would probably put the fraction of species which show some sort of polymorphism at between one-quarter and one-half. Most of those recorded are visual polymorphisms, not surprisingly because we are visual animals, and one might guess from this type of evidence that non-visual polymorphisms were equally common. Of those for which the genetics is known, by far the most common system is one where some alleles exhibit dominance over others, so that heterozygotes are not identifiable.

8.3 CONTINUOUSLY VARYING CHARACTERS

Continuously varying or quantitative characters tend to have normally distributed phenotypes. The genetic model developed for their control suggests that extreme variants should be homozygous at many of the controlling loci, while individuals near the mean have a higher degree of heterozygosity. Natural variation therefore implies underlying allele polymorphism at numerous loci.

It is also often the case that the characters will respond to selection. This implies the availability of a range of alleles to choose from. Figure 8.1 shows a summary of a classic experiment on selection by K. Mather and B.J. Harrison. It was published in 1949, and shows results typical of the experience of quantitative geneticists before and since. The experiment started with the selection for increasing number of abdominal bristles in *Drosophila melanogaster*. The population responded by an increase in number, and at the same time fertility fell. When selection was relaxed, the bristle number dropped and remained low, while the fertility increased again. A second high line derived from this stock increased in mean bristle number without loss of fertility, so that on relaxation of the selection the bristle number stayed high. Back selection from this high line resulted in a downward response, so that the high stock must still have carried decreasing genes. The experiment was used to show that bristle number and fertility are due to independent sets of genes, which can be moved around the genome by recombination. For the present purposes, it illustrates the following:

1. that organisms are phenotypically very plastic, implying genetic variability;
2. that both 'bristle number' genes and 'fertility' genes are present in the standard *Drosophila* population in a polymorphic state;
3. that it may be difficult to define the function of a given set of genes.

Since the loci affecting fertility initially also restricted the response to selection on the bristles they might be said to play a part in the control of

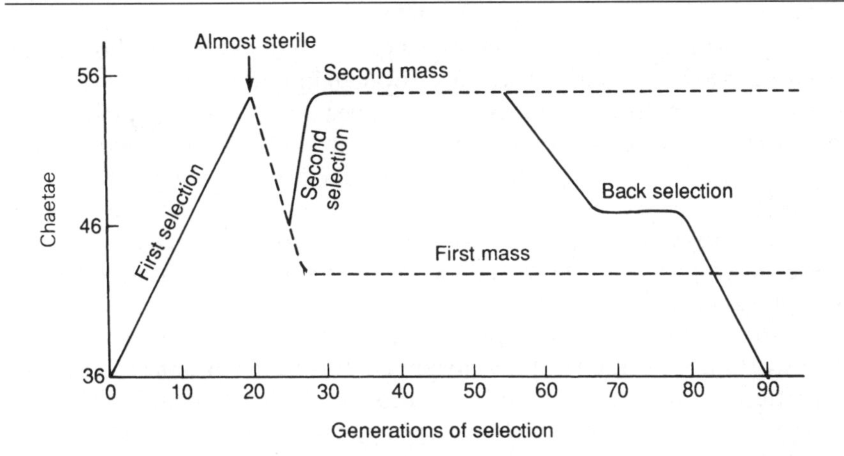

Fig. 8.1 Selection for increased number of abdominal chaetae in *Drosophila melanogaster*. In the first selection, increase in mean chaeta number was accompanied by a decrease in fertility. When selection for chaeta number was relaxed, fertility took charge and chaeta number fell. The second selection for chaeta number gave no correlated response in fertility, and the second relaxation of selection was followed by no fall in chaeta number, even though the stock was not homozygous as shown by successful back selection. Solid lines indicate selection and broken lines mass culture without selection. Data of Mather and Harrison, reproduced with the kind permission of Kluwer Academic Publishers from *Biometrical Genetics*, Mather and Jinks (1971).

bristle number. It is difficult to get an idea from this type of result how many loci are polymorphic. Since almost any quantitative character will respond to selection, however, it would appear to be a substantial fraction.

8.4 ENZYME AND MOLECULAR STUDIES

The genetic control of continuously varying characters is rather remote from the level at which we can visualize segregating alleles, and the evidence for them is indirect. Visual characters are much more dramatic evidence for polymorphism. They are picked up because they are obvious, however, and consequently there may be a bias towards overestimating the prevalence of genetic variability. In the early 1960s, two groups realized that enzyme polymorphisms detected by electrophoresis did not suffer from these drawbacks. Material from an animal or plant was put into solution and its constituents separated on a gel in the presence of an electric current. When stained selectively, bands indicating the presence of specific enzymes show up on the gel. Sometimes a single band indicates that the individual is homozygous for an allele at the locus concerned, while the presence of two or more bands indicates that it is heterozygous. This technique is limited only by skill in choosing gel, current and staining procedure suitable for detecting a given enzyme. It has the great virtue of picking up the primary products of the genes involved and of allowing all genotypes to be distinguished. An additional important bonus is that random samples of loci may be examined, since there is no way of knowing before the run which loci may be polymorphic. In 1966, R.C. Lewontin and J.L. Hubby working with *Drosophila*, and H. Harris working with human material, published papers in which they showed that random samples examined by electrophoresis indicated that about one-third of the loci were polymorphic. A new and powerful method of examining polymorphisms was opened up.

One of the results of this approach has been the adoption of a more general working definition of polymorphism. It is by no means obvious that mutation is not the cause of this variation, and a locus is said to be polymorphic if the most common allele has a frequency of less than 0.99. Many loci have been shown to segregate for several alleles, for which all the genotypes may be scored. The genotype frequencies do not usually differ greatly from the Hardy–Weinberg expectations. As a result, another parameter has come to be employed, the heterozygosity, H. This is the fraction of individuals in a sample which are heterozygous, calculated as $1 - \Sigma q_i^2$ where q_i is the frequency of the ith allele at a locus. An average value $\overline{H}$ may also be calculated by averaging over all loci studied. The original human data provided a value of 0.1, while the *Drosophila* experiment produced slightly higher estimates of $\overline{H}$.

Since then, thousands of studies have been carried out which add to the evidence of enzyme variability. They have been published independently by many different workers, but Eviatar Nevo has collected the results together in several surveys in order to see what pattern emerges. By 1984 he had data for 1111 species from the full range of animal and plant groups. The estimates of the fraction of loci polymorphic and the heterozygosity are summarized in Table 8.1. There are differences in mean values between groups, and when the data are analysed in more detail there is evidence of correlations with various environmental variables. For our present purposes, this summary table gives ample evidence of the widespread prevalence of polymorphism. The original observations are confirmed.

Enzyme variants detected by electrophoresis may not be typical of other classes of genetic material. Indeed, further work on technique indicates that the original methods underestimate the degree of variability present. It is also useful, however, to look at other types of genetic system. One of these examined by Lewontin was the evidence from blood groups. Blood groups were first detected because individuals of different genotype suffered antigen–antibody reactions when their blood was transfused, so naturally the first loci to be discovered were polymorphic. However, as serological techniques developed, more and more nearly monomorphic loci were discovered and the average heterozygosity and polymorphism decreased. Figure 8.2 shows the situation after these

Table 8.1 Average heterozygosity $\overline{H}$ and frequency of polymorphic loci, P, for a variety of studies collected together by Nevo *et al.*

	Species (no.)	Populations (no.)	Individuals (no.)	Average no. of loci studied	Average hetero-zygosity ($\overline{H}$)	Frequency of poly-morphic loci (P)
Mammals, excl. man	164	1112	28 653	24.4	0.050	0.222
Man	1	9	7349	107	0.125	0.470
Birds	41	97	2779	22.8	0.050	0.233
Reptiles	69	275	5152	22.1	0.052	0.240
Amphibians	60	448	9756	21.1	0.082	0.309
Fish	168	664	36 801	24.5	0.050	0.222
Echinoderms	15	19	815	20.4	0.109	0.479
Insects excl. Drosophila	116	385	22 406	20.5	0.077	0.316
Drosophila	33	383	29 776	26.8	0.115	0.419
Crustacea	116	298	18 195	23.2	0.091	0.342
Chelicerata	6	21	405	20.8	0.093	0.311
Molluscs excl. slugs	37	176	10 146	22.5	0.313	0.624
Slugs	5	28	1581	18.2	0.0	0.0
Nematodes	4	4	2249	24.0	0.014	0.076
Coelenterates	5	8	169	17.6	0.147	0.567
Flowering plants (monocotyledons)	5	141	3978	26.2	0.062	0.303
Flowering plants (dicotyledons)	39	820	19 095	19.4	0.059	0.311
Conifers	4	53	914	20.8	0.152	0.914

The numbers of species, populations, individuals and loci on which the studies are based are also shown to give an idea of the strength of the evidence. Within some categories listed there is heterogeneity between groups which have been averaged in this table, some of which is related to breeding pattern while other differences probably arise from differences in experimental technique. The figures nevertheless show the general levels of $\overline{H}$ and P attained. Nevo *et al.* analysed the variation in detail in relation to a number of ecological variables. Reproduced with the permission of Springer-Verlag from *Evolutionary Dynamics of Genetic Diversity* (ed. G.S. Mani), Nevo *et al.* (1984).

values have settled down to an apparently steady level. Once again the proportion of loci polymorphic is between 30% and 40%, and this figure is more likely to be an underestimate than an overestimate.

Molecular genetics has produced much more evidence of genetic variability. Detailed studies carried out on many genes, among them the *Adh* (alcohol dehydrogenase) locus in *Drosophila* and the *β-globin* locus in man, show that substantial levels of nucleotide polymorphism occur. Often the nucleotide differences detected produce synonymous codons, so that the gene product is unaffected. High levels of variation also occur in the non-coding introns. Restriction enzymes which act at the sites of sequences a few bases long are used to detect polymorphisms in the sequence lengths, called restriction fragment length polymorphisms (RFLPs). The numbers present in human material run into hundreds. Samples have individuals with sequences of different lengths because a restriction site recognized by the appropriate enzyme in one is absent in another. This failure of recognition indicates substitution of at least one base, so length polymorphisms can estimate the amount of variation at the nucleotide level. DNA fingerprinting uses the same technique to analyse intermediately repetitive nuclear DNA. The number of fragments, the number of repeats within fragments and the number of single-base mutations detected ensure that the probability that two individuals are identical is infinitesimal. In sequencing studies the frequency of polymorphism per nucleotide site is low, sometimes of the order of 1%, sometimes much less, but in a chain of sites the result is that many variant sequences will be present even when sequences are of quite moderate length.

Mitochondrial DNA (mtDNA) allows other interesting comparisons to be made. It is haploid and usually transmitted through eggs but not sperm. As a result it passes from generation to generation as if in a population only a quarter of the size of that carrying the nuclear genes, and the rate of dispersion is correspondingly greater. There is no opportunity for crossing-over, but the rate of mutation is higher than in nuclear genes. There are no repetitive sequences and a large amount of the sequence is transcribed. Because of these properties, mtDNA has been used

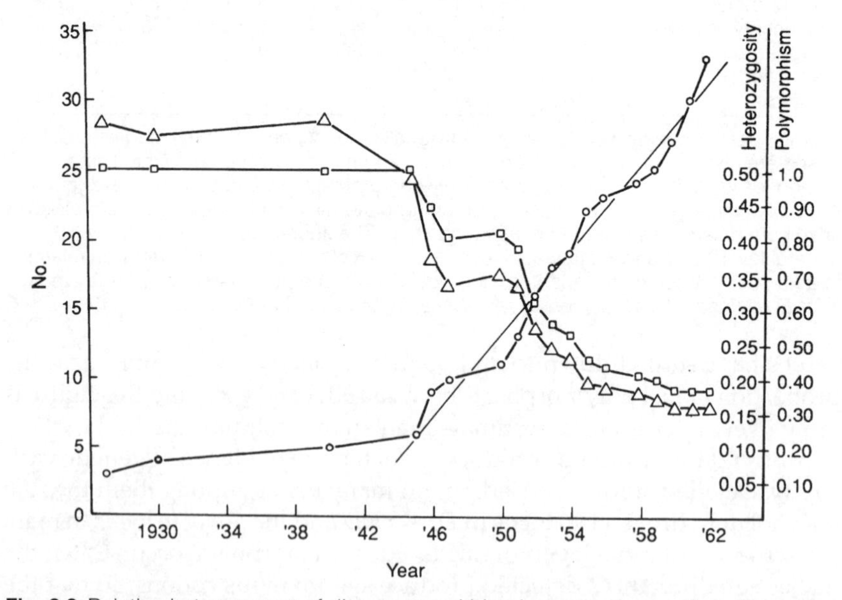

Fig. 8.2 Relation between year of discovery and blood group polymorphism. The curves show how the number of polymorphic loci and the heterozygosity have dropped to an apparently steady level as the number of loci discovered has increased. O, number of known loci; △, heterozygosity; □, proportion of loci polymorphic. Republished with permission of the Columbia University Press, New York. *The Genetic Basis of Evolutionary Change* (illustration), R.C. Lewontin (1974). Reproduced by permission of the publisher via Copyright Clearance Center, Inc.

particularly in studies of phylogeny, but polymorphism within populations is also revealed. As before, variants which produce synonymous codons are the most frequent. These lines of investigation therefore confirm other indications of high levels of genetic variation within populations. There is a higher level of polymorphism at the DNA level than at the level of the polypeptide chain. Introns and third codons are more likely to vary than first and second codons, and non-coding regions such as repetitive sequences and regions which could be coding but are untranscribed are also very variable.

8.5 CONCLUSION

All lines of evidence point to there being high levels of polymorphism in natural populations. Different types of study agree in suggesting that 30% of loci are polymorphic, while molecular evidence indicates that at the nucleic acid level the figure should probably be much higher. This variability allows populations to respond rapidly to new evolutionary opportunities, but it cannot be generated for that purpose. So what does generate and maintain polymorphism? This question leads on to almost all others in population genetics. We shall now examine some of the answers which have been proposed.

8.6 SUMMARY

Looking at the broad sweep of evolution, we see a process which appears to be guided by natural selection. Many of the changes which have taken place manifestly fit the species concerned for survival. It looks reasonable to generalize and to say that all variation is under the influence of natural selection. A consequence should be that populations are essentially monomorphic, since among the array of types potentially available one will be the fittest and the rest will tend to be eliminated. When genetic variation in natural populations is examined that is not what we observe; high levels of variability are apparently maintained. There are three ways out of this difficulty. First, selection may not have time to effect the elimination of unfit variants. This is unlikely if natural selection is strong because sufficient time is available for a low level of variability to be maintained. Second, appearances may be misleading and selection is not as pervasive as it seems. In that case, the variability may be generated by mutation and become fixed by random processes. Third, selection may be powerful but may act to maintain variability in populations. It is not self-evident why it should do so, and in that case we need to examine possible ways in which variation would be retained by selection.

8.7 FURTHER READING

Avise, W. (1987) *Molecular Markers, Natural History and Evolution.*
Berry, R.J. (1977) *Inheritance and Natural History.*
Cavalli-Sforza, L.L., Menozzi, P. and Piazza, A. (1994) *The History and Geography of Human Genes.*
Endler, J.A. (1986) *Natural Selection in the Wild.*
Ford, E.B. (1971) *Ecological Genetics.*
Lewontin, R.C. (1974) *The Genetic Basis of Evolutionary Change.*
Maynard Smith, J. (1998) *Evolutionary Genetics.*
Sheppard, P.M. (1975) *Natural Selection and Heredity.*

Polymorphism and ecological genetics 9

9.1 INTRODUCTION

One approach to understanding why polymorphism is so prevalent is to take a particular case and subject it to intense scrutiny. It is necessary to be a good naturalist or observer, to find a suitable example and then to build up a picture of the kind of forces operating. This approach was used in Britain by E.B. Ford and P.M. Sheppard, and in the USA by Theodosius Dobzhansky. The thinking involved is one of the threads which is woven into our current understanding of population genetics. The definition of polymorphism used is that of E.B. Ford (section 8.2), and polymorphisms are assumed to be present primarily as a result of selection. The only way a pair of alleles may be maintained at an equilibrium by selection, if the fitnesses of the genotypes are constant, is by heterozygote advantage (section 7.5.3). It therefore seemed probable that the selective forces, which may be varied and complex in their effect, balance each other in some way to produce heterozygote advantage. One example was strikingly successful in supporting this assumption.

9.2 HETEROZYGOTE ADVANTAGE AND TRANSIENT POLYMORPHISM

A number of variant forms of haemoglobin and other constituents of the blood are known, some of which have well-defined geographical distributions. The inherited disease sickle cell anaemia is widespread in Africa. It is caused by change in the haemoglobin, while a disease of the Mediterranean and the Middle East known as thalassaemia is caused by altering haemoglobin production. Haemoglobin F is found in South East Asia and Indonesia, while another disease, favism, caused by deficiency of the enzyme glucose 6-phosphate dehydrogenase (G6PDH), occurs in the Mediterranean and eastwards through Asia. The distribution more or less coincides with that of the severe malaria caused by *Plasmodium falciparum*, which was a major cause of childhood mortality before treatment and control measures were available. J.B.S. Haldane suggested that these genetic conditions, which are normally deleterious for their bearers, may also confer protection against the malarial disease.

The possibility has been investigated most thoroughly with respect to sickle cell anaemia. Homozygotes have haemoglobin S instead of haemoglobin A. This is a poor oxygen carrier which causes the red cells to become deformed under conditions of low oxygen tension. The result is anaemia, thromboses and susceptibility to a variety of diseases. Heterozygotes have both types of haemoglobin in their cells and the deleterious effects are small. There are 146 amino acids in the β chains of haemoglobin, and the difference between the two types is caused by substitution of valine for glutamic acid in position 6 of the chain. This results from conversion of the CTT or CTC sequences in the normal haemoglobin to CAT or CAC in the S haemoglobin. A single base substitution in a molecule, which is coded by 1722 bases in all, may have a profound effect on fitness. The S haemoglobin aggregates to form chains, causing the buckling of red cells which gives sickle cell anaemia its name, and the change in charge arising from substitution of valine for glutamic acid ensures that the two types may be distinguished by electrophoresis.

Haemoglobin C is another variant found in Africa, which has a story linked to that of haemoglobin S. It results from replacement of the glutamic acid at position 6 in the β chain by lysine. Again a single base change is required to cause the difference.

The difference between the molecules detected by electrophoresis allowed A.C. Allison to measure the frequencies of A and S genotypes in people in parts of Africa where malaria is prevalent. The genotypes were present in Hardy–Weinberg proportions among infants, but there were marked discrepancies among adults. The change allowed the fitness of the bearers of the three genotypes to be estimated as

AA	AS	SS
0.89	1	0.20

Section 7.5.3 explains how these values may be obtained from the change in frequency. The differences are due, on the one hand, to the anaemia and debility suffered by SS homozygotes, and on the other, to the fact that the heterozygotes are less susceptible to malaria than AA homozygotes. The malarial sporozoites get into the red blood cells, which carry both types of haemoglobin in heterozygotes. It is probable that these cells provide a worse environment for the malarial organisms, and also that such infected cells are likely to be destroyed by collapse within the bloodstream. In addition AS women are more fertile than AA women, probably because of a lower rate of abortion caused by malaria.

The AS polymorphism will therefore be stable as a result of heterozygote advantage as long as malaria is present. Frequencies are near the expected stable ones in East Africa, and have dropped in decendents of Africans taken to the New World and not now exposed to malaria.

In West Africa the situation is more complicated, due to the presence there of haemoglobin C. This is at a high frequency inland (8–12%) and drops as one moves southwards towards the coast while the frequency of haemoglobin S increases. Is this also a part of the equilibrium picture associated with malaria?

The S and C alleles are at the same locus, and for a three-allele system to be stable requires that all homozygotes have a fitness less than the mean fitness at equilibrium, or, which is the same thing, that at least two of the heterozygotes are fitter than all homozygotes and none is less fit than its associated homozygotes. (This may be proved theoretically by an extension of the type of reasoning in section 7.5.3.) The data have been examined by L.L. Cavalli-Sforza and W.F. Bodmer, who showed that estimates of fitness of the genotypes are

AA	AS	SS	SC	AC	CC
0.89	1	0.20	0.70	0.89	1.31

Examination of these figures shows that they do not meet the requirement for polymorphism. The CC genotype confers a higher fitness than any of the others. In the absence of C, we should expect stable polymorphism, but in its presence the result could be the take over of C, replacing both A and S. Of course, this outcome depends on the continued presence of malaria as an agency causing significant mortality, so that the ecology of the situation has always to be borne in mind. It also depends, not only on fitnesses, but on starting frequencies too.

If the frequency of A is p, that of S is q and that of C is r, then $\overline{w} = w_1p^2 + 2w_2pq + w_3q^2 + 2w_4qr + 2w_5pr + w_6r^2$. Given the fitness values obtained, we may find $\overline{w}$ for all possible combinations of the frequencies p, q and r of the three alleles. In Figure 9.1 the frequencies of A, S and C are shown along the edges of the triangle, while $\overline{w}$ is a surface within the triangle represented by a series of contours. Such surfaces (the 'adaptive topography' of the system) were discussed by Sewall Wright in relation to multi-locus models of interacting alleles. If r, the frequency of C, is zero, then there is a mean fitness peak of 0.903 at $p = 0.879$, $q = 0.121$. If C is present in the population, the highest peak, of 1.31, is at $r = 1$. Between these two values there is an 'adaptive valley', so that a population in which S appeared before C would move to the AS heterozygous peak, while one in which C appeared before S would move towards the CC peak. The former is probably what happened in practice. It is evident that the result is a population less well adapted (protected against malaria) than it would have been if all

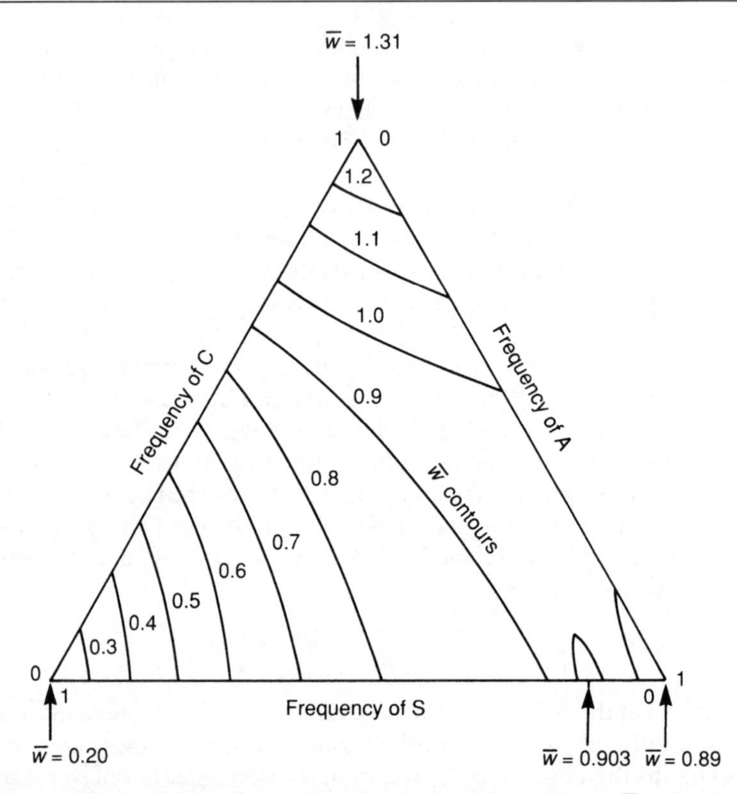

Fig. 9.1 An adaptive topography contour map of values of mean fitness $\overline{w}$ for the sickle cell and haemoglobin C system. Edges of the triangle show frequencies of the three alleles. The $\overline{w}$ surface is represented by contours. There are two low $\overline{w}$ corners to the surface, one high corner and a high point at the equilibrium frequency of A and S in the absence of C.

individuals had been CC. If the conditions of medical care and demography had not been changing, movement from the low to the high peak would only have taken place as a result of a mixture of selection and stochastic fluctuation in gene frequency.

The example therefore provides a case of stable polymorphism by heterozygote advantage, and a possible transient one. Both are driven by powerful selection. The other haemoglobin variants and the sex-linked G6PDH polymorphism have also been shown to confer protection against malaria. It is not clear whether these are examples of stable or transient polymorphism, but they fall into the same pattern as the ASC polymorphism. Studies on association of other human variants with disease have picked up some striking cases. The ABO blood group system and some of the loci in the HLA histocompatibility system affect disposition to disease.

9.3 CYANOGENESIS IN PLANTS

The Victorian hero General Charles Gordon was besieged in Khartoum by the Mahdi in 1884/85. As if there were not enough other problems, his horses began to die with symptoms of cyanide poisoning. Due to lack of forage, the horses had been feeding on leaves of the almond *Prunus amygdalus*. Almond leaves emit hydrogen cyanide when damaged and the smell of the gas is said to resemble that of bitter almonds. Another species of *Prunus* which releases hydrogen cyanide when its leaves are damaged is the cherry laurel *P. laurocerasus*; it used to be used by amateur entomologists for killing insects. As far as is known, all almond trees and cherry laurel bushes have the ability to produce hydrogen cyanide when their leaves are damaged. They are said to be cyanogenic and are monomorphic in that respect.

Other species of plants are polymorphic for the ability to emit hydrogen cyanide when damaged. Two examples are the bird's-foot trefoil *Lotus corniculatus* and white or Dutch clover *Trifolium repens*. They provide further illustration of the way alternative molecules may be selected, and broaden our view of the conditions which may lead to polymorphism. Two unlinked genes are involved, one coding for the substrate – a β-glucoside – and the other coding for the enzyme – a β-glucosidase – which degrades it. The situation in *T. repens* is especially interesting since European populations exhibit a complete spectrum of variation from fully cyanogenic populations in southern Spain to completely acyanogenic ones in Russia. Populations in intermediate locations show intermediate frequencies of cyanogenesis, changes in frequency corresponding with mean January isotherms (Figure 9.2). Two aspects of this polymorphism are informative. First, the frequencies of the alleles giving substrate and functional enzyme are significantly correlated. As these genes are on separate chromosomes the association indicates that they are being selected together. Second, the variation in frequency of cyanogenesis across Europe results from the interplay of two environmental factors of opposite effect on the polymorphism. In southern and western regions, cyanogenesis confers a selective advantage because it deters invertebrate grazers such as molluscs and grasshoppers. In continental and mountainous regions, such a potential advantage is more than offset by the increased frost sensitivity of cyanogenic plants so that cyanogenesis is rare. These opposing forces do not, in themselves, account for the polymorphism; in an isolated population it would be expected that one or the other would prevail. However, the selective grazing by invertebrates of acyanogenic plants at the seedling stage (Ennos, 1982) is possibly balanced by lower frost sensitivity. In addition, genotypes differing at the Li (linamarase) locus have distinct ecological niches. They are at an advantage in their own particular niches because they are more competitive, but

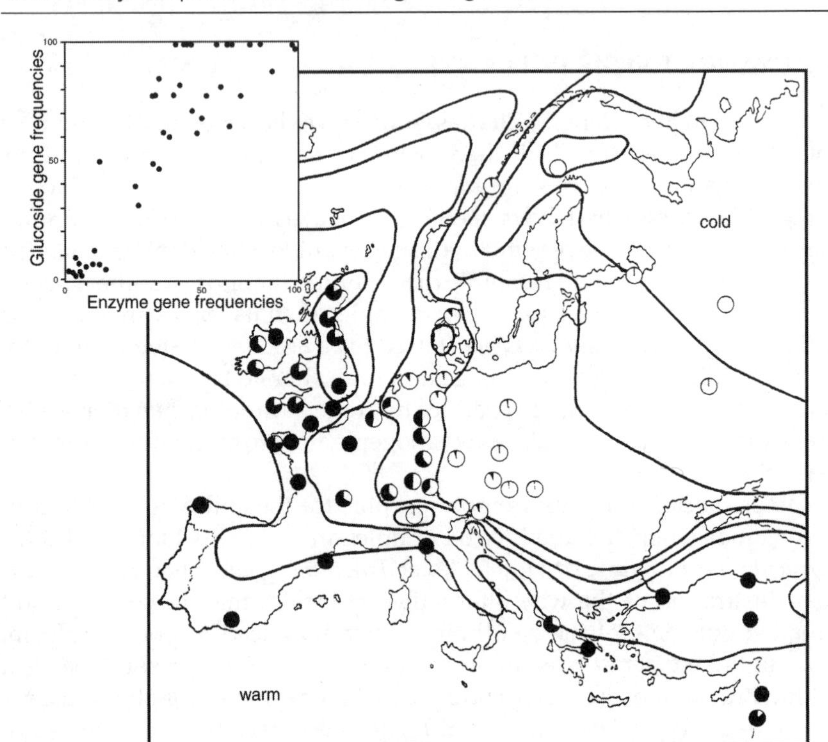

Fig. 9.2 Cyanogenesis in Dutch clover *Trifolium repens*. Distribution and frequency of cyanogenic glucoside gene in Europe in relation to winter temperature. Black sections of circles: glucoside allele frequency, isotherms: mean January temperature. Inset: relation of glucoside allele frequency (vertical axis) to enzyme allele frequency (horizontal axis) in the samples. Reproduced with the permission of Blackwell Science Ltd from *Heredity*, Daday (1954).

as its frequency increases, each genotype moves into alternative niches where it has lower fitness (Ennos, 1983).

9.4 SOME OTHER WAYS TO MAINTAIN POLYMORPHISM

Cepaea nemoralis is a gastropod snail which is common in western Europe and has also been introduced into the USA. It lives in a wide variety of habitats, including hedgerows, deciduous woodland and sand dunes. Another species, *C. hortensis*, is extremely similar in appearance, has slightly different habitat preferences but is often found in mixed colonies with it. The snails are hermaphrodites but obligate outbreeders. Both species are polymorphic for well-defined shell colour and pattern forms, and they have been the subject of the largest study of the maintenance of polymorphism in natural populations to have come from the ecological genetic school. There are three main shell colour forms, brown, pink and yellow. Within each there are two or three shades of colour. All are controlled by a series of alleles at a single locus, showing complete dominance of one allele over another. Brown is dominant to pink, which in turn is dominant to yellow, and darker shades are dominant to paler ones. Closely linked to the colour locus is one which controls the presence or absence of five brown bands running round the shell. Absence of bands is dominant to five-banded. There are other linked genes which modify the appearance of the bands. One, punctate, changes what is usually a continuous brown stripe into a series of spots. Another, hyalozonate, removes all or part of the pigment in the bands so that they appear more or less translucent. It also has an allele which removes the brown pigment from the lip of the shell in normally dark-lipped *C. nemoralis*, or causes its deposition in normally white-lipped *C. hortensis*. Apart from lip colour, these loci can only manifest themselves in an individual homozygous for five-banded at the unbanded locus, i.e. there is epistatic interaction. Two loci, unlinked to this group or to each other, also modify the banding pattern. One of them, trifasciate, removes the top two of the five bands, so that only the lower part of each whorl is banded. The other, midbanded, removes the top and the bottom two, so that there is a single equatorial band on the shell. There is also non-segregating variation in number of bands and in fusions between them. Hundreds of colonies have been sampled, from all types of habitat occupied by the snails, and almost without exception the colonies are polymorphic. The polymorphism has the virtue that evidence of it remains on the shell after the animal is dead, and it is known from fossil samples that the variability has always been present. Many other snails, including related ones in the family Helicidae, are more or less invariant in shell colour and pattern, so what kind of explanation may be advanced in this case?

The first type of explanation for the diversity was that the shell colour and pattern are not selected. If all the different combinations have equal fitness, then frequencies could fluctuate at random from place to place, and in the 1930s this was as an acceptable interpretation.

However, more evidence about the pattern in the field with some thought and reference to the theory in section 7.4, shows that drift on its own will not do. Random fluctuation in gene frequency leads inexorably to fixation, so that although many different alleles may be present in the species as a whole they should occur in monomorphic patches of an area determined by the neighbourhood size, or N_e. In fact, frequencies change from place to place and combinations of alleles replace one another, but even tiny colonies are polymorphic. Some force must be maintaining the variant forms.

The problem was discussed in these terms in 1951 by the French population geneticist M. Lamotte. In Chapter 7 the different forces affecting gene frequency are discussed separately one at a time. In practice they operate together, and if systematic forces tend to move gene frequency towards an equilibrium then a dynamic balance will be set up between them and the drift which leads to gene dispersion. The result will be a distribution of frequencies in populations, with a modal frequency and a variance determined by the size of the systematic forces and by N_e. The smaller N_e the larger the variance and the more spread out the distribution. Sewall Wright derived an equation for describing such a stationary gene frequency distribution. Lamotte had examined a large number of populations in the field and recorded their morph frequencies, so that he was able to draw empirical gene frequency distributions. In the course of this work, he observed that there was evidence of selection in some places, but none that selection operated to produce a balance. He therefore considered the possibility that forward and back mutation might do so instead. Using the Wright formula and an estimate of N_e of 500, which accords with the experience of most *Cepaea* workers, he estimated the size of the mutation rates which would have to be involved. It turned out that they were larger by two or more orders of magnitude than most mutation rates. Polymorphism can only be explained simply by a balance of drift and mutation if the loci are especially mutable. Part of the systematic component could be contributed by migration, however, so that given the right circumstances it is possible to conceive a model where different selection in different populations with migration between them ends up ensuring that all the populations are polymorphic, and we will consider later what this model might be like.

An intensive investigation of the force of natural selection was carried out by A.J. Cain and P.M. Sheppard in England. They noted that in the areas of agricultural land and woodland where they worked the general colour of the snails tended to match that of the background. The match was by no means exact, but dark and uniform places such as beech woods had populations which were largely brown or pink, rather than yellow, and unbanded rather than banded. Pale and diverse habitats such as hedgerows, on the other hand, supported yellow-banded snails. This background matching was the result of visual selection by song thrushes, which feed intensively on the snails at certain times of the year. Cain and Sheppard observed the predation taking place and were able to show that the morph frequencies of the victims, as indicated by the remaining shells, differed from those of the populations from which they came in having a higher frequency of the more conspicuous types. The selection pressures could be large, 5–10%, and applied to phenotypes rather than to genes.

Some populations contain pink unbandeds and yellow bandeds at high frequency, because those are the inconspicuous combinations, rather than proportional combinations of the alleles at the different loci. The selection therefore generates linkage disequilibrium. Particular chromosomes, or haplotypes, are favoured and the linked loci come to operate as if they were a single locus of multiple alleles (pink unbanded, yellow five-banded, etc.), instead of independent entities. Such a tight combination of linked loci with linkage disequilibrium is sometimes called a super-gene. These observations showed that in seeking an explanation of the polymorphism one had to look to selection, including visual selection, and to interpret the phenotype of an individual rather than the separate effects of the alleles at the different loci.

But how does the selection work? No balance of forces is known, akin to that operating on the AS haemoglobin polymorphism, to produce heterozygote advantage. That does not necessarily mean it is not there. Its detection is made difficult by the dominance in visual appearance of the morphs, so that heterozygotes cannot be compared with homozygotes, and by the difficulty of knowing what other pleiotropic effects of the genes may be involved. Experiments and surveys suggest that the morphs may be tolerant of extremes of heat and cold to different extents, yellow unbanded being the most resistant form. However, no theory has been built up showing how this would balance the visual selection. The absence of evidence may indicate that heterozygote advantage does not occur, so that it is worth considering other possibilities.

B.C. Clarke examined the way in which predators behave when faced with a choice of prey. There is plenty of evidence from a variety of kinds of observation and experiment that vertebrate predators act in such a way that they take a larger fraction of common forms, and consequently a smaller fraction of rare forms, than is available. Some examples from invertebrate predators have also been demonstrated. There are two kinds of explanation for this. One is that the predators optimize their feeding behaviour in such a way as to take a larger fraction of common forms. The other is that the brain works in such a way that rare opportunities are underexploited. In either case the result for the prey species is that rare mutants will have an advantage and increase in frequency. There is evidence from the distribution of morphs in *Cepaea* populations that the predation exerted by song thrushes may work in this way. Clarke described the result of this behaviour as apostatic selection, and pointed out that it favours distinctly different phenotypes at the expense of common ones. In *Cepaea* there is evidence of background matching, but none of the morphs is a particularly good match to any specific element of the background, so that the range of forms present is consistent with the idea that distinct forms are being favoured. The best example of a field study on molluscs showing evidence of apostatic selection comes from a polymorphic snail, *Littoraria filosa*, living on mangrove trees (Figure 9.3). Such a direct demonstration is not available in *Cepaea*, but there is evidence for the right kind of predator behaviour. The polymorphism could therefore result from apostatic selection. To operate in this way, the selection has to vary with the frequency of the phenotype, not the gene. Represented formally, selective values would have to have the following form.

PP	PY	YY
$w_1f(p)$	$w_1f(p)$	$w_3f(q)$

Here, it is assumed that there are two phenotypes, pink and yellow, with pink dominant to yellow. The fitness of each genotype making up this pair of phenotypes is represented as a constant (w_1 or w_3) plus a part which varies with p or q, respectively. The function $f(p)$ has to be large compared with $f(q)$ when pink is rare, and vice versa. The selection is therefore genotype- or phenotype-frequency dependent. One of the important outcomes of the study of *Cepaea*, and other examples which were examined at the same time, is the observation that selection may often be dependent on commoness of the morph, rather than independent of it. If this is so, heterozygote advantage is not a necessary requisite for polymorphism.

Another type of balance which may contribute to the *Cepaea* polymorphism was mentioned in connection with the investigation of Lamotte. Suppose there is selection in different directions in different adjacent

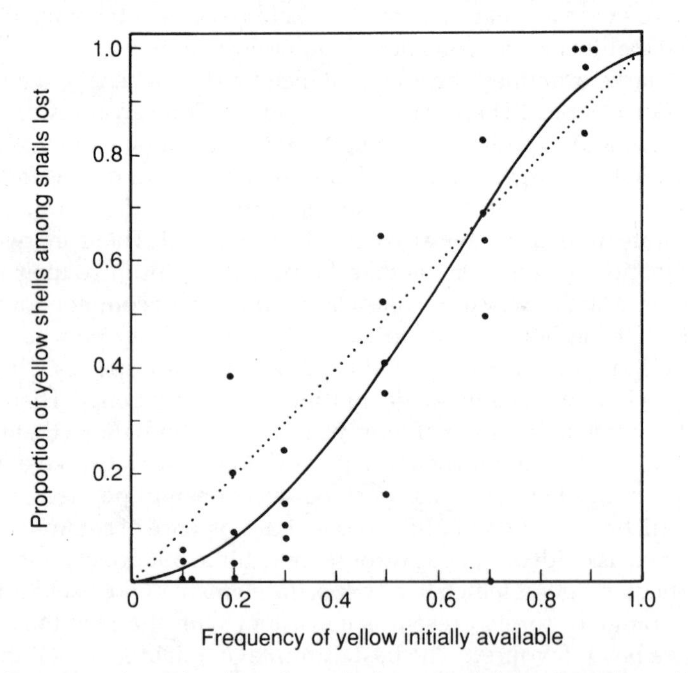

Fig. 9.3 Frequency-dependent loss of morphs from populations of a polymorphic snail. The mangrove snail *Littoraria filosa* has yellow and brown morphs. Small bushes of the genus *Avicennia* were colonized with different frequencies of yellows. The graph shows the fraction of yellows among individuals lost after 2 weeks plotted on the fraction originally present. The straight line is the relation expected if individuals were lost in proportion to their original frequency. The solid curve is fitted to show the pattern followed by the data. When yellows are infrequent, they are lost less than proportionally, when very common they are lost slightly more than proportionally. It is assumed, although not demonstrated, that the result is due to predation. From Reid (1987).

populations of *Cepaea*. A polymorphism could be maintained if there was migration between them and the selection was so balanced that the exchange counteracted the tendency for loss of an allele in either population. A model of this type, perhaps with exchange between a whole network of populations and a more haphazard array of selective values, may be visualized.

The simplest situation would involve two populations with migration between them. If the advantage is small, the disadvantage must also be small, so that the balance must be very precise, although when selective values are larger the balance is less restrictive. If we consider a population divided into two subunits c_1 and c_2, then the selective values which will balance each other also depend on the relative size of c_1 and c_2, so that for this reason too, the model is not robust. Theoretical considerations therefore lead us to be cautious about models of this type as a general explanation for polymorphism. Their applicability depends, however, on what kind of natural phenomenon is being explained. If we consider a network of demes, instead of just two, with differing selective values and migration between them, then it becomes very likely that at least some of the demes will be polymorphic, so that it is necessary to decide whether this pattern fits the field data.

Formally, such models require that each of the populations is self-limiting, so that excess individuals are forced into suboptimal environments. The model is therefore similar to the last one, except that the selection now changes with density rather than with frequency. In either case, the selective values are variable and depend on the commoness of a morph.

Equilibrium can also arise through fluctuation in selection with time. Selection favouring a particular morph could proceed for a number of generations, to be followed by selection against it. Again, a precise balance is required for stable polymorphism. Two-unit models of this type may be represented as in Figure 9.4. The fitness of a recessive morph is shown on the ordinate as W_d, while the fitness of the same morph at another time or place is represented on the abcissa as W_a. It is necessary that the fitness values fall between the two lines. The lower, straight line in Figure 9.4(a) is $W_d = 2 - W_a$, and it is necessary for population values to be above it. This is equivalent to saying that the arithmetic mean of W_a and W_d must be greater than 1. For varying fitnesses fluctuating with time, the upper, curved line indicates that the geometric mean of the fitnesses must be less than 1. If fitness varies in space, the curved line will be higher, and represents the harmonic mean of the fitness. When there is both spatial heterogeneity and habitat selection on the part of the individuals, this range is again somewhat widened. It is difficult to judge how often these conditions might be met, but heterozygote advantage is by no means the only way in which selection could maintain a polymorphism.

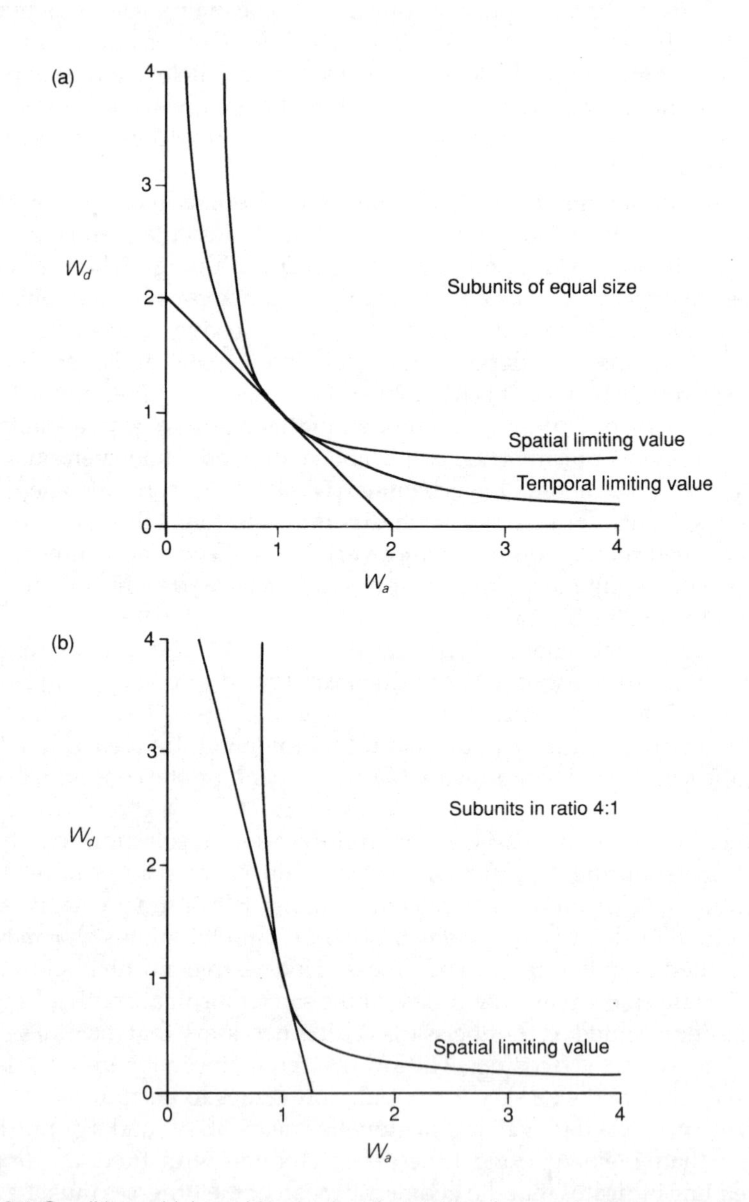

Fig. 9.4 Conditions which would maintain polymorphism for two morphs in an environment fluctuating in time or space. The fitness of the recessive morph must fall between the straight and the curved line. The position of the straight line in (a) indicates that the arithmetic mean of the fitnesses must exceed 1. The upper lines in (a) indicate that the geometric mean of the fitnesses must be less than 1, for temporal heterogeneity, or that the harmonic mean must be less than 1 for spatial heterogeneity. In (a) the spatial model assumes equality of subunits. In (b) they are in the ratio 4:1, so that balancing selective values are asymmetrical.

In accounting for the polymorphism in particular populations, another problem is to know the timescale which should be used. The explanations which have been discussed relate to theoretically stable polymorphisms, but if the species is relatively long-lived in relation to the observer then what is in fact a transitory phase may look stable. In *Cepaea nemoralis* there are localities in southern England where frequencies of particular morphs are quite invariant. These adjoin other areas where the morph frequencies are constant about different mean values. A.J. Cain and J.D. Currey termed the patterns observed 'area effects', to draw attention to them. One type of explanation which they considered was that the morph frequency patches mapped on to environmental patches, perhaps of some factor like the quality of the soil, which determined the frequencies and which had not been independently detected. An alternative explanation is that different founder individuals colonized the different areas, and as they increased in numbers and extent, grew so different in genetic constitution that the sharp transitions where their descendants met were due to hybrid inviability between the colonies of the different areas. Such a partial separation could arise if fitness depends to an important extent on interaction between alleles at different loci. This possibility is considered further in Chapter 16 and a general model for *Cepaea* polymorphism in Chapter 14.

9.5 IS HETEROZYGOTE ADVANTAGE GENERAL?

As ecological genetic studies have accumulated, so more and more evidence of strong selection has become available. Phenotypes are often controlled by super-genes, and research may be expected to unfold the action of selection on numerous pleiotropic effects of the combinations of alleles concerned. Experience suggests, however, that frequency- or density-dependent selection is commonplace. Since it is powerful compared with mutation, it must be involved in maintaining the polymorphism when it occurs. What is not obvious is that the result usually adds up to heterozygote advantage; at present that remains unproved. If it should turn out to be the usual explanation for this type of polymorphism then we may reasonably ask why the heterozygote should so often have elevated fitness. An inference from the ecological genetic approach is that the advantage may evolve.

9.6 EVOLUTION OF HETEROZYGOTE ADVANTAGE

In *Trifolium repens* the two loci controlling cyanogenesis are unlinked. Selection favours alternative pairs of alleles to produce disequilibrium but segregation redistributes them in each generation. Such a situation would ultimately favour any mechanism which restricts recombination between the genes showing linkage disequilibrium, provided that other advantageous combinations are not destroyed in the process. Genes on separate chromosomes can be protected from recombination by interchange of non-homologous chromosomal segments (see Chapter 13). The loci controlling the shell colour and pattern morphs of *Cepaea* form a group of closely linked loci in linkage disequilibrium, i.e. what E.B. Ford called a super-gene. Numerous species offering other examples may be cited, some of the most graphic being those controlling the mimetic polymorphisms of butterflies. C.A. Clarke and P.M. Sheppard have studied several examples. Thus, the swallowtail butterfly *Papilio memnon* mimics a number of tailless distasteful butterfly models (Figure 9.5). In order to switch from one form to another, there have to be changes in presence or absence of tails, the hindwing pattern, the forewing pattern, the colour of the basal triangle on the forewings and the abdomen colour. Breeding experiments show that these may vary independently but that they are controlled by a group of linked loci so that the change from one combination to another is made in concert. The combination forms a super-gene with extreme linkage disequilibrium. The butterfly is widespread in South East Asia. Within part of its range, the different combinations behave as allelic series showing dominance, but if crosses which would not normally occur are made between areas, the dominance breaks down and intermediates appear. These observations suggest that a high degree of integration of the system controlling the polymorphism has evolved.

The explanation of how it could evolve goes back to some early work of R.A. Fisher. He suggested that because a heterozygote produces half as much of each of two gene products, it should, on average, be intermediate in expression between the homozygotes. Nevertheless, most mutants which we observe are recessive. Fisher explained this by saying that since mutants are deleterious most of the time and are usually present as heterozygotes, selection will favour any change in the genome which tends to make them recessive. We have seen that dominance and recessiveness are descriptions of the expression of genes, and that genetic variance in expression may be detected (sections 5.3 and 5.5). The change in expression must be due to genes at other loci which modify the genes studied, and over very long periods of time the appropriate modifiers have accumulated to render the average mutant recessive.

Applying this argument to alleles which are pleiotropic but have a net advantage, the implication is that advantageous attributes should in due course become dominant and disadvantageous ones recessive. What follows is most easily seen by attaching the model to an example. The typical form of the peppered moth, *Biston betularia*, was at one time the most prevalent form over the whole of Britain, and is well camouflaged on trees in rural and non-industrial areas. As urbanization and industrialization increased in the nineteenth century, a melanic mutant, *carbonaria*, increased in frequency in areas subject to atmospheric pollution where it was the most cryptic form. The increase in frequency appears to be due to differential predation of the conspicuous individuals. There is some evidence from examination of nineteenth century collections that heterozygotes may at first have been intermediate in appearance between the homozygous forms, and it is reasonable to suggest that production of large amounts of melanin carries with it some sort of metabolic disadvantage. At the start of the selective process, the two fitnesses could be represented like this:

genotype	TT	TC	CC	
first component	1	$1 + \frac{1}{2}s$	$1 + s$	as a result of visual selection
second component	1	$1 - \frac{1}{2}t$	$1 - t$	as a result of non-visual selection

The mutant increased in frequency because S has a larger value than t. Application of the modification of dominance argument suggests that the advantageous effect should become dominant and the disadvantageous one recessive. If the changes were brought about successfully, the fitnesses would become:

TT	TC	CC
1	$1 + s$	$1 + s$
1	1	$1 - t$

The result is that the TC heterozygote now has the highest fitness, so that instead of the melanic replacing the typical the system ends in the state of balanced polymorphism. There is no evidence that this actually has happened in the case of the peppered moth, but the example gives a clear illustration of the argument.

The model provides a possible selective explanation for polymorphism. Not only should heterozygote advantage evolve but we should also find groups of linked loci concerned with the same system built into a super-gene. That is because any mechanism reducing recombination between the loci involved would reduce the fraction of unfit offspring, and be favoured as a result. Some such systems exist, such as the shell colour and pattern polymorphisms in snails or the mimicry polymorphisms in butterflies, and the expression scored usually shows complete dominance. That is not to say that they evolved in the way described, however, and the proposal attracted strong criticism from the moment it was published.

Two types of criticism may be advanced. On the one hand, there are other ways in which the systems may arise. It has repeatedly been argued that the kinetics of enzyme production may make dominance of heterozygotes more likely than an intermediate expression, and linked groups of genes with similar functions may arise, not by a process of selection for linkage but by non-reciprocal crossing-over. This is likely to be the origin of the globin multi-gene families, for example.

The other objection is statistical. If the frequency of alleles at the modifier loci is to be changed by selection affecting the expression of alleles at another locus, then the effect on the modifiers must inevitably be small. The modifier alleles have to be effectively neutral, and even then a change will not come about unless the modifier loci are closely linked to the major locus from the start. These arguments have not been resolved conclusively, and new discoveries in genetics may have changed the terms of reference within which the problems can be framed. The theory of the evolution of dominance and of heterozygote advantage remains a fascinating proposal, which accounts for some of the phenomena we see associated with polymorphism and may explain how they arose.

9.7 SUMMARY

The study of natural history shows that there is a high incidence of genetic variability in the wild. Much of this consists of the coexistence of discontinuous phenotypes in populations at levels which could not be maintained by deleterious mutations (the genetic polymorphism of Ford). The conclusion is that polymorphism, and probably genetic variation in general, even when it does not manifest itself as polymorphism, is maintained by a balance of selective forces.

If the selection acting on the genotypes is constant then stable equilibria indicate heterozygote advantage. Some examples are known where heterozygote advantage has definitely been demonstrated, but not many, and it is possible that the widespread polymorphism is a consequence of the prevalence of frequency- and density-dependent selection. This could arise as a result of frequency-sensitive patterns of predation or of balance between selection in different directions at different times or in different parts of the habitat. At a larger scale, it could result from balance of selection and of long-distance migration.

The chance that heterozygote advantage will arise is enhanced by the fact that different pleiotropic effects of a gene may differ in their dominance. Recurrent deleterious mutations tend to be recessive, which would be explained by selection over long periods of time for reduction of their expression in heterozygotes (Fisher's theory of the evolution of dominance). Using the same argument for a polymorphic locus, if advantageous effects tend to become dominant and disadvantageous effects recessive then heterozygote advantage will evolve (the Sheppard–Caspari theory). Consequently, populations may evolve to become more polymorphic. There is evidence that expression can be altered experimentally, although the genetic nature of the modifying process is unclear.

Fig. 9.5 Mimicry in swallowtail butterflies. The subfamily Papilioninae of the family Papilionidae contains three tribes. One of these, the Troidini, consists of distasteful insects, the larvae feeding on poisonous plants containing alkaloids, which are retained in the adult insects. The Papilionini feed on plants lacking the alkaloids, and are edible to predators (Linnaeus and other early taxonomists named many of the distasteful species after the Trojan and the edible ones after the Hellenic protagonists in the Homeric legend). Throughout the world there are associations of troidine models and papilionine mimics. The models are monomorphic and usually invariant throughout their range while the mimetic females are frequently polytypic, mimicking different models in different parts of the range, and they are sometimes polymorphic. Where the ranges of model and mimic do not overlap completely, that of the model is larger than that of the mimic. In Africa, some females of *Papilio dardanus* resemble the distasteful danaid butterfly *Danaus chrysippus*, others resemble another danaid, *Amauris niavius*, and others the acraeid *Bematistes poggei*.

Fig. 9.5 (continued) The illustrations here show polytypy in *Papilio memnon*. This species is widely distributed over the South East Asian archipelagos and resembles different models on different islands. Many swallowtail butterflies have similar elements to the patterns of their wings, not least the tails themselves. Mimicry is not simply achieved by resemblance of homologous components, however, but also involves different features of the model and mimic. *Parides coon*, *Pachlioptera aristolochiae* and *Parides sycorax* have bright areas, red or white on the head and thorax. In *P. memnon*, these are mimicked by red or white patches on the base of the forewings. Similarity depends on: (a) presence or absence of tails; (b) colour and pattern of the hindwings; (c) colour and pattern of the forewings; (d) colour of the basal triangle on the forewings, and (e) abdomen colour. Breeding experiments have shown that in sympatric crosses between variant individuals alleles controlling each component exhibit dominance, but that the dominance tends to break down in allopatric crosses. Analysis of rare varieties resulting from recombination suggests that the pattern is controlled by a super-gene in which the loci are arranged in the sequence shown above. The difference in palatability between model and mimic species, the coincidence of model and mimic patterns on different islands, the existence of a super-gene composed of functionally unlike loci and the dependence of dominance on sympatry all indicate that the mimetic resemblance has been developed, and is not a consequence of close phylogenetic relationship. This evidence, the ratios of model to mimic found in natural communities and direct observation of predator behaviour show that the selection is imposed by visually hunting predators. The systematics and relationships of the papilionid species are discussed by Munroe (1961) and the genetics of *P. memnon* by Clarke and Sheppard in papers in 1971 and earlier.

The study of highly organized polymorphisms, such as those determining alternative mimetic types in butterflies, reveals the existence of sets of linked loci responsible for related aspects of the phenotype. These super-genes have evolved as a result of selection for complementary combinations of alleles at different loci. They suggest that to some extent the genome has evolved to become a set of groups of linked loci showing high levels of linkage disequilibrium.

9.8 FURTHER READING

Berry, R.J. (1977) *Inheritance and Natural History.*
Ford, E.B. (1971) *Ecological Genetics.*
Majerus, M.E.N. (1998) *Melanism. Evolution in Action.*
Sheppard, P.M. (1975) *Natural Selection and Heredity.*

Response to selection by pesticides 10

10.1 THE PROBLEM OF RESISTANCE

Pesticides are essential to control insects, mites and nematodes which attack crops and are vectors of diseases. Evidence of genetic resistance to pesticides has become increasingly apparent, however, and application of population genetics may help in understanding the patterns of selection involved, and allow strategies to minimize the effect of resistance to be devised. Hundreds of cases of resistant species have been recorded over the past 40 years, indicating response times on the part of the animals of between 10 and a few hundred generations. By the early 1960s, the use of insecticides resulted in a drop of 30% in the world total of deaths from diseases carried by arthropod vectors. In the 1970s, there was a resurgence in several important diseases; malaria had declined to an annual number of cases of around 70 million, and then increased to over 200 million. Not all of this change is due to resistance, but the recorded incidence of resistance has paralleled the increase in malaria. On the agricultural side, approximately 20% of the world's crop is lost to pests, and it is estimated that without pesticides this figure would be increased to about one-half. There is therefore a lot at stake, both in terms of health and agricultural produce and in terms of money.

One way of dealing with the problem is to introduce new pesticides. This can be a very effective approach. Modern pesticides are environmentally less harmful than those first used. A few years ago, however, it was estimated that the cost of development of a new product was around $10 million, with a development time of about 7 years, so that it is impossible by this means to keep pace with the response of the target organisms. Resistance is, in effect, an added recurrent cost of pesticides, so that as time goes on they also become more and more expensive to apply.

There are four main classes of chemicals used as pesticides. These are organochlorines such as DDT and dieldrin, organophosphates such as malathion and parathion, carbamates (e.g. carbaryl) and pyrethroids (e.g. permethrin). Pyrethroids are synthetic pyrethrum-like substances, and pyrethrum is a naturally occurring plant chemical, long used as an insecticide, which probably evolved as a defence against insects in the first place. In some cases, the insects may have had a long time to evolve responses to the natural analogues of the synthetic products in current use.

The types of response observed have been classified into four groups. First, the organism may develop a behavioural response which results in its avoiding the pesticide, by increased negative reaction to sublethal doses or preference for untreated microhabitats. Second, there may be decreased cuticular penetration when the insect comes into contact with the chemical. If that fails, metabolic pathways may be developed which detoxify the pesticide once it is in the animal. Finally, target organs, often the nervous system, may become less sensitive to the effects of the poison.

Generally, the mode of action is similar within one of the pesticide classes, so that there is likely to be cross-resistance between chemically related products. This reduces the number of pesticides effectively available for long-term application. In addition, cases of multiple resistance are known, where an insect has developed genetically distinct mechanisms of resistance in response to different insecticide classes. A well-documented case of failure of control concerns house flies on Danish farms, which have been subject to intensive insecticide control programmes for the past 40 years. Resistance to products in all groups is now recorded, with many cases of cross-resistance, and a time to loss of control by a new product of 1–3 years. This situation is probably representative of that in any country where intensive application takes place. We have to continue to use chemical control methods, and population genetics can provide no sweeping solution to the problem. It has an important part to play, however, if it can supply some rules which allow strategies to be developed to minimize the chance of loss of control.

10.2 PATTERN OF SELECTION

When a pesticide is first applied to a population of insects, it kills those individuals it comes into contact with and produces a large reduction in numbers. Any individuals of resistant genotypes which are in the population, or enter it by mutation or migration, are relatively more likely than susceptibles to survive and produce offspring, so that in the course of time the frequency of susceptibles decreases and the population size rises. As a population genetic system, resistance at a single locus illustrates the pattern of strong selection, leading in principle to replacement of one allele by another and progressive decrease of the initial high genetic load. The processes are described in the preceding sections, the only special feature of pesticide resistance concerns the dominance of the alleles involved.

The pattern of mortality often observed in laboratory tests of single loci conferring resistance to pesticides is shown in Figure 5.1. The total picture is built up from a series of independent tests of mortality at different doses. Critical doses can then be found which allow the three genotypes to be distinguished.

In the field a particular concentration of pesticide is applied, which may render the resistance gene effectively recessive or effectively dominant. If insects come into contact with low concentrations, heterozygotes and homozygotes will survive, but if the concentration is high, only the homozygotes will survive. At least to some extent, the dominance of the resistance gene is under the control of the investigator. Now we have seen that at low frequencies a recessive gene takes many more generations to increase in frequency under selection than a dominant one (Figure 7.5). This suggests that the pesticide should be used as sparingly as possible in order to minimize the chance that resistance will develop, but that when it is used the concentration should be high, so that the susceptible is rendered effectively dominant and the resistant allele recessive.

Investigation of some mosquito populations in Africa showed that they were probably polymorphic for dieldrin resistance (albeit with low frequencies of resistance genes) when dieldrin was first used. Natural resistance is not unexpected if a pesticide resembles a naturally occurring plant substance in structure. However, when a new pesticide is first used, resistance genes will usually be rare. When use of the pesticide ceases, the frequency of resistance has been observed to drop. This is not surprising, because the alleles are likely to have other effects which reduce fitness in the absence of the pesticide. Increase in resistance in populations subject to treatment therefore occurs because the protection conferred by the resistance gene outweighs the deleterious pleiotropic effects.

The net advantage of carriers of the resistance genes is a product of the relative fitness of resistant and susceptible in the presence of the pesticide, and of the probability of coming into contact with it. Typically, the environment is complex and will offer refuges from the pesticide, into which the insects may escape. These may be particular breeding sites in mosquitoes, diapause stages in moths, alternative food plants in aphids or resting, as opposed to feeding, sites in house flies. In the refugia, the frequency of resistance will decline, so that a judicious programme of treatment also involves understanding the ecology of the target insect sufficiently well to allow untreated refugia of appropriate size to exist. On the larger scale, migrants into a treated area from untreated neighbouring localities will bring in susceptible individuals, especially if treatment has resulted in a drastic reduction in numbers. This will be seen as a drawback to anyone trying to exterminate the pest, but may have advantages from the point of view of ensuring that resistance does not become a problem in the longer term. The following simple model illustrates some of the patterns of change in frequency of resistant alleles which may be deduced from these considerations.

10.3 A GENERAL MODEL

Suppose a population consists of insects carrying susceptible alleles for response to a particular insecticide at an autosomal locus at frequency q'. The resistant alleles are initially at some low frequency $p' = 1 - q'$. Generations are non-overlapping. The population is large and in each generation random mating brings the genotype frequencies to near the Hardy–Weinberg frequencies. Mutation may be represented by writing $q = q' - uq'$, the rate of mutation from resistant to susceptible being ignored. We will assume that resistance may confer some kind of disadvantage on carriers of the resistance allele, and for simplicity, that the disadvantage is recessive and acts in the pre-adult stage. Adults from elsewhere enter the population, along with those emerging *in situ*. The frequency of fresh adults of the three genotypes is

$$a = \frac{p^2(1-m)}{\overline{w}} + mp_m^2$$

$$b = \frac{2pq(1-s)(1-m)}{\overline{w}} + 2mp_m q_m$$

$$c = \frac{q^2(1-s)(1-m)}{\overline{w}} + mq_m^2$$

where m is the migration rate and $\overline{w} = p^2 + (1-s)(1-p^2)$.

Insecticide is now applied to the adult population. The fitness of the susceptible compared with the resistant homozygote is $1 - r$, and that of the heterozygote is $1 - hr$. The factor h varies from 0 to 1 to determine the dominance of the resistance gene, and is a function of the concentration of insecticide with which an insect comes into contact. If the concentration is low, h is near zero, and it increases to 1 as the concentration goes up. Individuals have a probability e of escaping altogether, so that the three genotype frequencies become

$$a' = \frac{a}{\overline{w}}$$

$$b' = \frac{b(1-hr)(1-e)+be}{\overline{w}}$$

$$c' = \frac{c(1-r)(1-e)+ce}{\overline{w}}$$

where $\overline{w}$ is the summation of the new terms. We can now calculate q_1 as $c' + \frac{1}{2}b'$, and go through the cycle again to get q_2, etc.

It has been assumed that to a reasonable approximation the immigrants are in Hardy–Weinberg frequencies when they arrive, but this may not be so. Movement may occur at other stages of the life cycle, and the disadvantageous effect of resistance may not be present or may not be recessive. For any particular case, all these steps would have to be checked against the ecology of the population; deviations from the assumptions may affect the outcome considerably.

We can now look at some of the consequences of varying the parameters. Figure 10.1(a) shows the decline in frequency of the susceptible gene from the high initial frequency of 0.9999. Mutation is at a rate of 10^{-5}, there is no immigration or escape, and the resistance gene is not disadvantageous in the absence of insecticide. When resistance is dominant ($h = 0$, insecticide concentration low) and $r = 0.2$, a frequency of $q = 0.5$, which

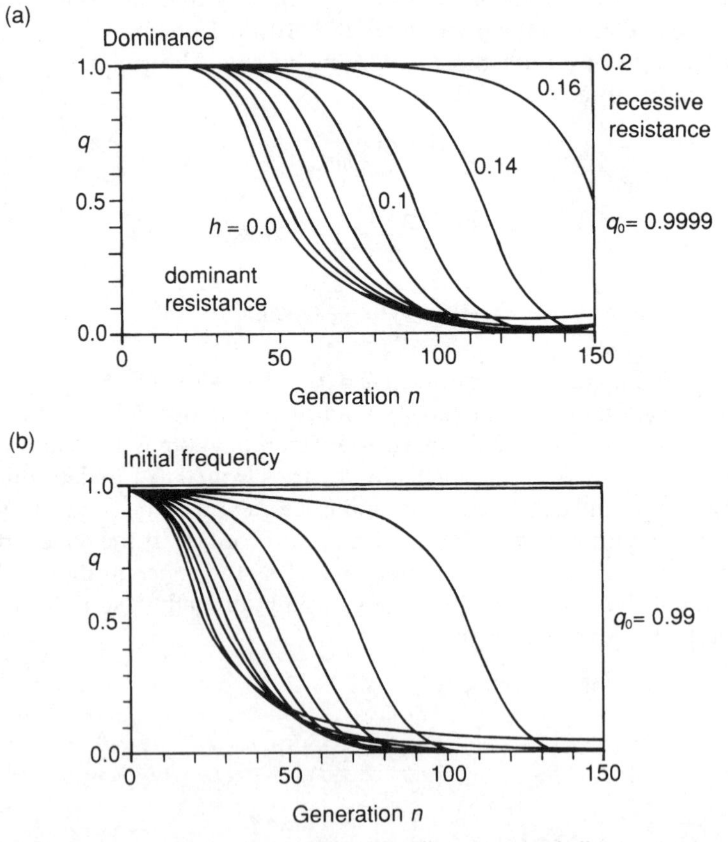

Fig. 10.1 Response to selection by insecticide at different levels of effective dominance of the resistant allele. Gene frequency of the susceptible allele is plotted on number of generations. The susceptible homozygote is 20% less fit that the resistant homozygote. The resistant gene is dominant when $h = 0$ and fully recessive when $h = 1$. The initial frequency of susceptibles is 0.9999 in (a) and 0.99 in (b). Other details are given in the text.

implies loss of control of numbers, is reached after less than 50 generations, but this frequency is not reached in 150 generations when $h = 0.18$ or more. Clearly, insecticide concentration is an important factor to consider in practice. Figure 10.1(b) illustrates an identical set of conditions except that the initial susceptible gene frequency is reduced to 0.99. All responses take place in about half the time. In the field, these two initial gene frequencies would be impossible to distinguish.

The effects of escape and of immigration from susceptible neighbouring populations are examined in Figure 10.2. Here, the initial susceptible frequency is 0.99, mutation rate is 10^{-5} and $h = 0$. In Figure 10.2(a) the fraction escaping varies from 0 to 100%, while in Figure 10.2(b) immigration varies from 0 to 20%. At worst, loss of control due to increase of resistants takes 25 generations, while an 80% escape level or a few per cent

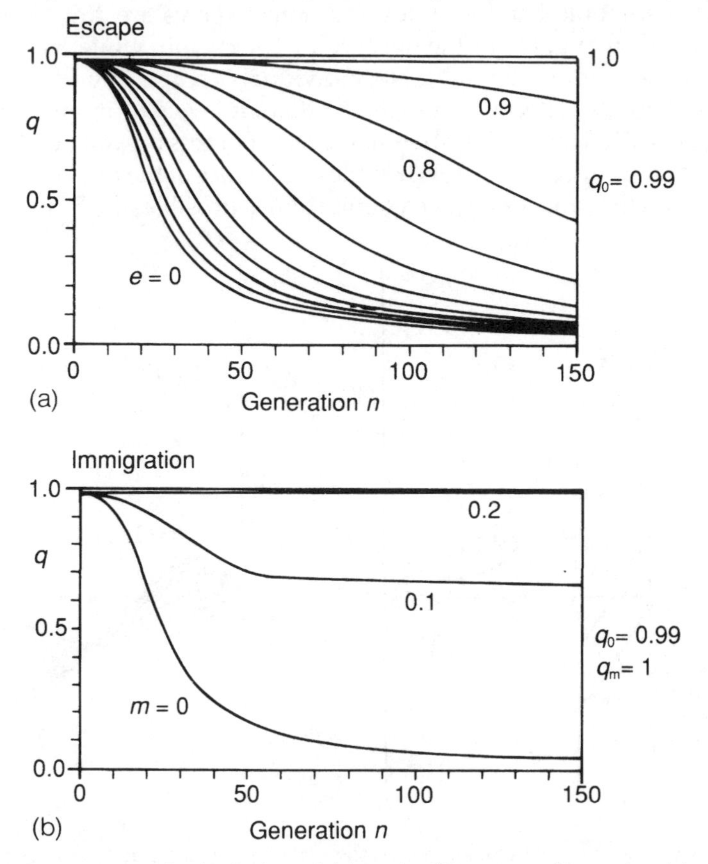

Fig. 10.2 Response to selection by insecticide under different conditions of avoidance and of immigration. The left-hand curve is the same as the left-hand curve in Figure 10.1(b). (a) shows the effect of progressively larger fractions of the population escaping into refugia where they are unaffected by the insecticide. In (b) the response is diluted by immigration of individuals from a susceptible population.

susceptible immigrants put this point off indefinitely. Success in this respect, however, means survival of relatively large numbers of the pest insects, and the effect on immigration depends on the dynamics of the populations from which the immigrants come, which are themselves receiving resistant individuals from the population studied.

Polymorphism may arise in several ways. If selection in favour of resistant individuals is balanced by influx of susceptible immigrants then the population could settle down to an equilibrium level, at least for a period of time. If the innate fitness of the resistant allele has a different dominance from the fitness component conferred by the insecticide then the net result could be heterozygote advantage. These two possibilities are illustrated in Figure 10.3. The graphs show the relation of Δq to q, a negative path crossing the line $\Delta q = 0$ indicating a stable equilibrium. In Figure 10.3(a) susceptible heterozygotes and homozygotes are 25% fitter than resistant homozygotes in the absence of insecticide, while its presence reduces their fitness by 20%. No individuals escape into refugia. The curves show the effect of a range of different levels of immigration by susceptibles. In Figure 10.3(b) similar levels of fitness prevail but the dominance of resistance changes. When the dominance of the two components of fitness is the same ($h = 1$), no equilibrium is possible, the fitness values

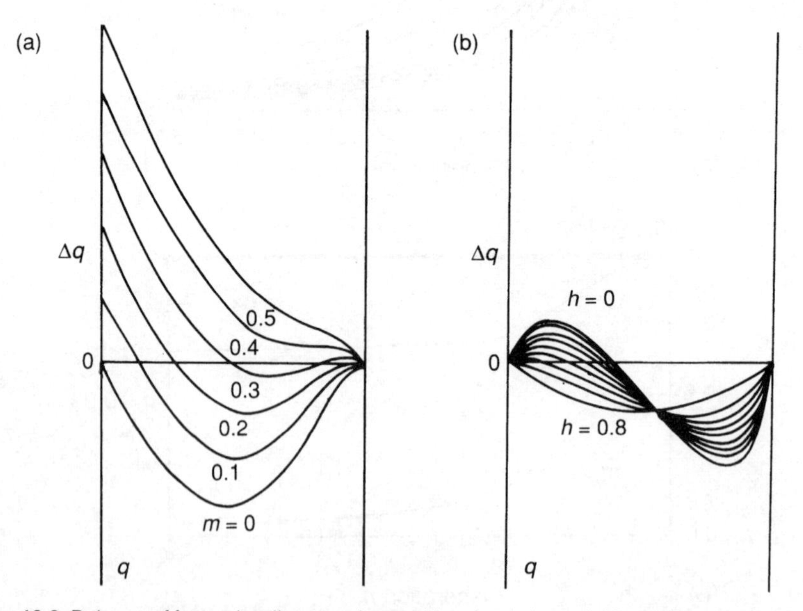

Fig. 10.3 Balance of forces leading to polymorphism for an insecticide-resistance allele. In (a) selection of the resistant allele is balanced by immigration of susceptibles. In (b) resistant homozygotes have an innate disadvantage compared with heterozygotes and susceptible homozygotes. This is balanced by the effect of the insecticide on fitness when this effect is dominant (*h* near 0) but not when it is recessive (*h* near 1). Other details are given in the text.

chosen resulting in loss of the susceptible allele. As h moves towards zero, the system moves to stable polymorphism.

In reality, all the factors discussed operate together, and a random drift component also accompanies them. As a result of mutation and migration resistant genes may well enter a population soon after it is treated with insecticide. Heavy localized application in the presence of refugia and susceptible immigrants will reduce the chance of their frequency increasing to the point where control is lost. In plants, resistance to herbicides is much less prevalent than insecticide resistance in insects, and it has been suggested that this is because of a phenomenon akin to the immigration effect. The new growth after treatment by a herbicide includes many plants from the untreated seed bank present in the soil, as well as direct descendants of the treated generation, and these unselected plants reduce the chance that resistance will develop. Herbivorous insects exhibit resistance more commonly than their predators or parasitoids. One reason could be that destruction of the herbivores is followed by an overshoot in the number of herbivores but a delay in the build-up of predator number. The contribution of susceptible immigrants is therefore relatively more important in the predator population than in the herbivore population. Tests of field data on pests of fruit trees suggest that this is the reason in their case. The alternative is that herbivorous insects are pre-adapted because their food plants contain insecticide-like protective agents to which they have already responded, so that there is an element of cross-resistance.

These considerations lead to the possibility that application of mixtures of insecticides, or of different insecticides in a mosaic of patches, may be beneficial. If two insecticides are used in the same place, alleles resistant to one are likely to be lost because the carrier is susceptible to the other. If they are used in neighbouring patches then immigrants are unselected for the resistance genes selected in each patch. Understanding of this kind of system requires that linkage distance between the loci, and linkage disequilibrium, be considered. Both courses of action may be useful, but whether they are or not depends very much on the particular circumstances of the trial. There is no simple remedy for the resistance problem, but theoretical exploration indicates the type of ecological and genetic parameters which must be estimated and some possible guidelines for the intelligent use of pesticides. Regular routine monitoring and appropriate patterns of application are essential if we are to avoid repeated loss of control. Their costs should be incorporated into the total cost of pest control programmes.

10.4 SUMMARY

Pesticide resistance is a response by natural populations to selection. It is medically and economically important, and since the selection is strong its investigation favours the ecological genetic approach. The expression of resistance genes may depend on the concentration of pesticide experienced, sometimes varying from dominance at low to recessiveness at high concentrations. Resistance to new pesticides develops rapidly and often, and there is frequently cross-resistance between related products. The chance that some individuals will avoid exposure to any given application and that susceptibles from untreated areas will migrate to areas treated with pesticides make it likely that polymorphisms for resistance will be set up. At the same time, these considerations make it possible to devise strategies to minimize the likelihood that resistance will develop. In general, targetted individuals should be treated with a high concentration of pesticide, an untargetted reservoir of untreated susceptibles should be allowed to survive and treated plots should be surrounded by areas which are untreated or treated with unrelated pesticides. Although these procedures may be inconvenient, the ease with which populations can adapt and the cost of new chemical formulations make it prudent to adopt carefully thought out control programmes and to maintain adequate monitoring of resistance levels in pest species.

10.5 FURTHER READING

Georghiou, G.P. and Saito, T. (eds) (1983) *Pest Resistance to Pesticides*.

National Research Council (1986) *Pesticide Resistance. Strategies and Tactics for Management*.

Population consequence of selection 11

11.1 GENETIC LOAD

In places where the sickle cell polymorphism occurs at equilibrium, some individuals of SS type die of anaemia, and others of AA type die of malaria, who would not do so if they were fortunate enough to possess the AS genotype. We can work out the fraction of the population involved. Compared with the heterozygotes, the AA individuals have a fitness $(1 - s_1)$ of 0.89, so that $s_1 = 0.11$. The SS individuals have a fitness $(1 - s_3)$ of 0.20, so that $s_3 = 0.80$. These figures are given in section 9.2. In section 7.5.3 the equilibrium gene frequency for a system such as the sickle cell polymorphism is shown to be

$$\hat{q} = \frac{s_1}{s_1 + s_3}$$

which has a numerical value of 0.1209, so that $\hat{p} = 0.8791$. At equilibrium p^2 of the population suffers 11% loss through the effects of malaria, i.e. 8.5% are lost. Also, q^2 of the population suffer 80% loss. This comes to 1.17%. The total loss is 9.7%, which is the value, for the sickle cell polymorphism, of what is called the genetic load. This is related to the intensity discussed in section 4.8, and may be represented as

$$\frac{w_0 - \overline{w}}{w_0}$$

where w_0 is the fitness of the optimum genotype, or $1 - \overline{w}$ if $w_0 = 1$. In this example the presence of the polymorphism carries with it the implication that 10% of the population will be lost because they are homozygous for one allele or the other. Genetic load from this cause is called the segregational load.

11.2 MUTATIONAL LOAD

At the time of the atmospheric testing of nuclear bombs, H.J. Muller drew attention to the possible consequences of an artificially increased mutation rate. Deleterious mutants are kept at a low frequency by selective elimination, but what sort of frequency do they attain? The answer depends on the balance of mutation and selection. Suppose a locus mutates to a deleterious recessive form at rate u per generation, and that the reverse rate is effectively zero. If q represents the frequency of the mutant ($=1-p$) then the rate of mutation is up. Selection acts against these mutants, but only when they are present as homozygotes. Using the equations in section 7.5.3 it may be shown that for a recessive allele at frequency q the change in frequency due to selection is

$$\Delta q = \frac{-spq^2}{1-sq^2}$$

When q is very small the denominator is very nearly unity, so that Δq is nearly $-spq^2$.

Equilibrium will therefore be achieved when $up = -spq^2$, that is, when $u = sq^2$, or $q = \sqrt{u/s}$. This shows that if the selection against a recessive mutant is not very strong the mutant may not be uncommon. If the mutation rate is 10^{-6} and there is 1% selection against it the equilibrium frequency is 1%. The presence of the mutant imposes a load of sq^2, resulting from selection. But $q^2 = u/s$. Substituting u/s for q^2, we see that the load always has a value of u, the mutation rate for any locus at which there are deleterious mutants, whatever the selection against them. Muller described this as the mutational load. Now u is very small but applies independently to each locus. Since there are thousands of loci, the total effect is much larger and may run to several per cent. An increase in the mutation rate could therefore produce a very real increase in the loss from a population.

11.3 SUBSTITUTIONAL LOAD

Another substantial loss could occur as a result of the replacement of one allele by another. Even though the replacing allele is advantageous, we may sometimes wonder if the population can withstand the effects of selection. When the typical form of the peppered moth was being replaced by the melanic, its fitness was estimated to be only two-thirds of that of the favoured type. When the typical allele was at a frequency of 70%, 49% of individuals were typical homozygotes and one-third of those, amounting to 16% of the population, were destroyed per generation. Haldane described this loss as the cost of selection, and argued that few loci could undergo gene replacement at any time because of the limitations imposed by the high cost. It is also called substitutional load.

11.4 DOES LOAD LIMIT SELECTION?

The three types of genetic load referred to are aspects of a problem identified in sections 4.7 and 4.8. The stronger the selection, the smaller is the selected band of survivors. We may reasonably ask whether the high levels of genetic diversity seen in populations can be maintained by selection. This was one of the questions raised by Lewontin and Hubby when they demonstrated that about 30% of the genome in *Drosophila* was polymorphic, i.e. about 2000 loci. If each polymorphism were maintained by heterozygote advantage imposing its own segregation load, the implication is that the average level of selection would have to be minute for the species to survive at all. On average, the gene frequencies for each polymorphic locus will be distributed about 0.5, so that selection against each of the two homozygotes is the same. For each locus, the average size of the load is $s(p^2 + q^2)$, or $\frac{1}{2}s$. The fraction which is not lost in the selective process is $1 - \frac{1}{2}s$. If two loci are maintained, the average surviving is $(1 - \frac{1}{2}s)^2$, and for n loci it is $(1 - \frac{1}{2}s)^n$. This total must be well above zero if the population is to survive. If 50% survive segregational load then $(1 - \frac{1}{2}s)^n = 0.5$. With n equal to 2000 this means that s cannot be larger than 0.0007, which leaves little room for the levels of selection encountered in the sickle cell example or in *Cepaea* or the peppered moth. For many species this figure is probably of the order of $1/2N_e$, so that the average allele is effectively neutral. Strong selection goes hand in hand with high levels of mortality, and there is little chance that many polymorphisms could be maintained by selective balance.

But is this really correct? The implication is that much, perhaps most, of the mortality experienced by a population is associated with suboptimal phenotypes, and that output can only just compensate for the mortality. For many animals and plants, the problem seems rather to be how to check their exuberant growth, so that load is not seen as a problem. It has been suggested that in human populations the mutational load causes a large fraction of the early abortions and failures of implantation. If that is the case then the effect is hidden from sight and taken up in the excess of potential over realized fecundity. Even so, another look at the sickle cell example shows that the very formulation of the load problem may be couched in misleading terms.

The sickle cell polymorphism is undoubtedly maintained by differential mortality at the post-fetal stage. The load calculation estimates this mortality, which would not exist if all individuals had the AS genotype. However, it is important to compare the polymorphic state, not with the ideal of an all-AS population but with the all-AA population which must have been the reality before the S allele was present. If we do that we see that the existence of the polymorphism has actually reduced mortality (from 11% to 9.7%). Intuitively, that is what we should expect – the difficulty arises from the definition of mean fitness used in the calculation of load, which takes account only of frequency and not of the number of individuals in the different classes and the change in demography which has accompanied the arrival of the S allele. There are other reasons, as well, why the load calculation might be misleading.

11.5 OBJECTIONS TO THE LOAD ARGUMENT

Reservations about the load calculations may be summarized under headings 11.5.1 to 11.5.3 below.

11.5.1 THE CONCEPT OF OPTIMUM FITNESS

The load is calculated as $(w_0 - \overline{w})/w_0$. Algebraically, the logical choice for w_0 would be the highest fitness in the set which is being considered. For the sickle cell polymorphism, this is the fitness of the AS genotype. As we have seen above, however, it would be ecologically more realistic to choose the fitness of the AA genotype, in which case the load becomes negative. By the same argument, when we do the calculation for the 2000 loci in *Drosophila*, the implicit assumption is that the ideal genotype is the heterozygote for all 2000 loci. This has an infinitesimal probability of occurrence and no one knows what such a super-fly would really be like. If something more realistic is chosen for the optimum fitness, such as the average fitness of the fittest 10% of a population, the load is no longer enormous and larger selective values per locus can be seen to be sustainable.

11.5.2 LOCUS INTERACTION

Another assumption of the argument presented is that selection on each locus is independent of all others and the effect accumulates from locus to locus. This is certainly not always the case. Some unfit individuals are eliminated because they are homozygous at numerous loci, so that removal of a single individual may contribute to the maintenance of several polymorphisms. The quantitative genetic model indicates that the tails of the phenotypic distribution of a character are brought about by highly homozygous combinations of genes, so that stabilizing or threshold selection operates in this efficient way. If non-independence were taken into account, calculated values of load would be much lower.

11.5.3 SOFT SELECTION

Bruce Wallace introduced the terms hard and soft selection to distinguish between different effects on the population. Hard selection is density independent. Many deleterious genes will reduce the fitness of their carriers whatever the circumstances; hence their presence always leads to selective elimination. For others, however, the severity of the effect is a function of the prevailing conditions. In vials, *Drosophila* pupate on the sides in a band above the medium. As an example of soft selection Wallace supposed that the vial would support 25 pupae. Two types were present, type A which always survived up to the total and type B, which would only survive if space was left for it by type A. Selection therefore depends on the density and composition of the population. If type A were absent, type B could fill all the available space. If there were 25 type A individuals, type B could occupy none of it, and the fitness of B relative to A varies between these extremes. The selection of B depends on the number of A, but it can be represented in terms of relative frequency.

This model can be extended to a situation where there are two or more niches, in each of which one of the types has an advantage. An early example of discussion of frequency-dependent selection was provided by Dobzhansky and Wright. They studied two chromosome types of *Drosophila pseudoobscura* in cages in which the medium was continually renewed. The types compared were called ST and CH, and although they are in fact different chromosome inversions, they can be treated for the present purpose as if they were a pair of alleles at a single locus. Experiments with different starting frequencies showed that ST tended to increase from a low frequency, and to decrease from a high one, towards an equilibrium of about 0.7. When curves were fitted to the data, it was shown that for the system

	CH,CH	CH,ST	ST,ST
starting at	p^2	$2pq$	q^2

the fitness values on the basis of heterozygote advantage were

0.32	1	0.71

Constant fitness is assumed, and at equilibrium frequency there would be a segregational load of 0.203. This is calculated in the same way as the load for the sickle cell system. However, an equally good fit was obtained by using the fitness values

$$0.1 + 1.29q \qquad 1 \qquad 1.9 - 1.29q$$

where q is the gene frequency.

Here, fitness is frequency dependent, each genotype having an advantage when rare which decreases with increase in frequency. The pattern of selection is shown in Figure 11.1. This type of selection could arise if the environment contained different niches of limited extent, in each of which

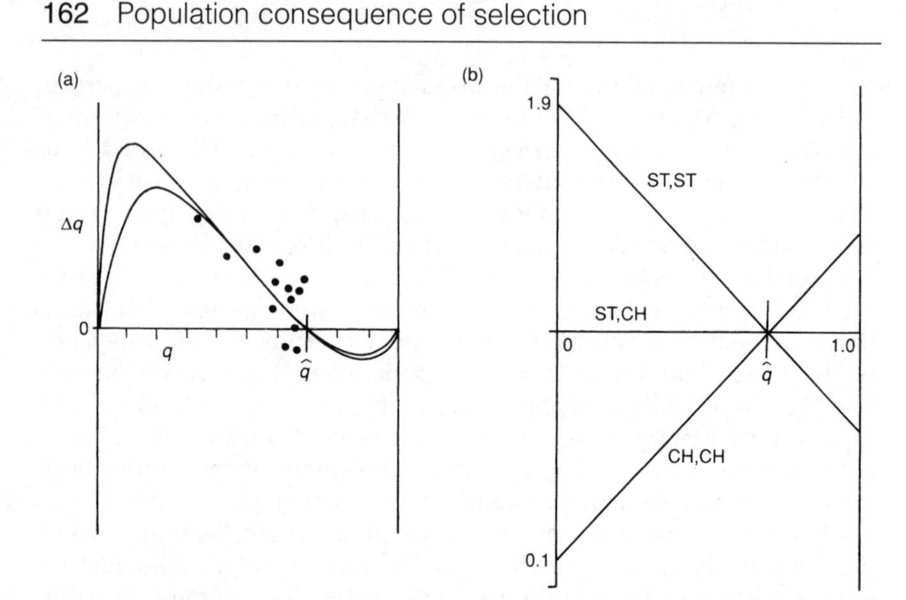

Fig. 11.1 Data from a classic experiment by Sewall Wright and Th. Dobzhansky, published in the journal *Genetics* in 1946. In it, they suggested that polymorphism for inversions in *Drosophila pseudoobscura* might be accounted for by frequency-dependent selection. This was one of the first experiments in which gene frequency changes were studied in artificial population cages. In (a) the relation of Δq to q is shown, where q is the frequency of the inversion ST. The lower curve is the one which would fit the data if there were heterozygote advantage; the upper one is for frequency-dependent selection. For the latter, the relation of the relative selective values to q is shown in (b).

either CH or ST was at a competitive advantage. The situation is discussed in different terms in section 9.4. For the present purpose, the important point is that the fitnesses change as the frequency approaches equilibrium, so that at equilibrium the load disappears. It will only be sustained if the frequency wanders as a result of drift and selection is required to restore the equilibrium. If a large amount of selection is density or frequency dependent, as appears to be the case, then maintenance of polymorphism may not entail a load.

11.6 CONCLUSION

For the reasons outlined above, the genetic load is likely to be much lower than early estimates indicated. To some extent the apparent problem was an artefact of the way in which selection is represented. Selection does not necessarily act independently on each locus, and the total loss cannot be estimated knowing only the relative fitness of the genotypes. Nevertheless, the argument is by no means entirely spurious. Selection must involve some sacrifice, either of actual or of potential recruits to a population, and strong selection will have a greater effect per unit time than weak. The problem is to estimate how serious the consequences are for a population. The supposition that mutational plus segregational load would impose an insupportable burden was one of the reasons for exploring the possibility that most polymorphisms are in fact neutral. The argument for neutrality is examined in the next chapter.

11.7 SUMMARY

Selection in favour of one kind of individual usually implies selection against another. The fraction of a population lost in the course of selection is called genetic load. In principle, it sets an upper limit to the amount of selection which can take place. An objection to the idea that most polymorphism is selectively balanced is that the implied load is too great to be sustained. It has also been argued that deleterious mutations and replacement of one allele by another impose severe constraint. The effect of load is not usually apparent, however. This may be partly because potential output is very high. Frequency- and density-dependent (soft) selection, and selection operating on several loci together also reduce the amount of genetic load. Consideration of the load argument does, however, suggest that heterozygote advantage operating independently is not the explanation for most polymorphisms.

11.8 FURTHER READING

Crow, J.F. and Kimura, M. (1970) *An Introduction to Population Genetics Theory.*
Wallace, B. (1970) *Genetic Load.*
Wallace, B. (1991) *Fifty Years of Genetic Load. An Odyssey.*

Polymorphism and neutral mutation 12

12.1 INTRODUCTION

Faced with the information that the ecological genetic approach did not reveal the expected heterozygote advantage, the arguments for genetic load and the rapidly developing field of molecular genetics, some people in the 1960s started to investigate the proposition that the polymorphisms may not be selected at all. There has always been a stream of debate in population genetics about the role of random drift in changing gene frequencies, but the new phase got off to a memorable start in 1969 with a paper in *Science* by J.L. King and T.H. Jukes entitled 'Non-Darwinian Evolution'. The argument has been developed to its fullest extent by Motoo Kimura and Masatoshi Nei. There are a number of reasons why some, perhaps most, mutations might not be subject to selection and it is possible to estimate the likelihood that they will be polymorphic.

12.2 ARGUMENTS FOR NEUTRAL MUTATION

A neutral mutation is one which has the same effect as another allele at the same locus, so that they are functionally equivalent. In practice, allele frequency may change as a result of drift, and selection is only detectable if it has an effect on frequency which can be distinguished from drift. Operationally, a neutral allele is one which has a selection coefficient of $1/2N_e$ or less. The following three types of evidence are in favour of neutral alleles.

The genetic code is redundant. There are 61 codons specifying amino acids. Each base can mutate in three different ways, so that a single-base substitution within a codon could give rise to nine different new forms. There are therefore $9 \times 61 = 549$ possible single-base substitutions, of which 134 are synonymous. We therefore have 134/549 possible mutations, i.e. 24%, which will not alter the polypeptide coded, and may therefore be assumed to be selectively neutral.

Another argument of a similar kind looks at the relative frequency of different amino acids. If one starts with the relative amounts of the four different bases, one can ask whether some are called upon more than others in the composition of amino acids. The method is to combine the bases into triplets at random in the proportions in which they are present. When this is done, the result is to produce relative frequencies of amino acids which, with the exception of arginine, are close to those actually observed. One interpretation of this is that polypeptides actually are made up of the products of random permutation of bases through evolutionary time.

Other evidence comes from examining the degree of differentiation between species. When the amino acid composition of their proteins is compared with the differentiation at the DNA level, it is found that the DNA has the higher divergence rate. There are several reasons for this. Third codons are found to be more variable than first or second codons. Introns are more variable than exons, and in the non-coding regions, repetitive sequences consist of units which differ between different taxa. The evolution of repetitive sequences has its own dynamics within a species or population, which often leads to predominance of particular sequences. Factors affecting the pattern of elements within a particular lineage are not subject to the same forces as the proteins controlled by the coding regions of DNA, and they may well be, to all intents, neutral. The same applies to variation which produces synonymous codons or which is in sequences excised before final construction of the polypeptides.

12.3 THE NEUTRAL THEORY

The main elements of the neutral theory are as follows. Organisms consist of a phenotype, which is controlled by the interaction of genes and environmental effects. Because of the complexity of the developmental process, there is a loose connectedness between the genetic and the phenotypic level, so that a number of different genomes, represented by the DNA sequences, can produce equivalent phenotypes. At the phenotypic level, selection works generally in a 'purifying' way to remove unfit individuals, and this selection may in fact be very strong. Its net result is in the direction of monomorphism. At the level of the DNA or of the primary polypeptides coded by the structural genes, however, there are many functionally equivalent variants. These are what are known as neutral alleles, and their distribution can be understood if we know the effective population size and the mutation rate. Polymorphism is evidence of variability which is *not* eliminated by selection, because selection does not touch it.

Depending as it does on only two parameters, the neutral model allows predictions to be set up as to the number of alleles to be expected, the mean and the variance of allele frequency distributions, and so on. These may be tested against data, and in this respect the theory has an advantage over models involving selection, which often do not lead to clear-cut predictions.

12.4 PREDICTIONS ARISING FROM THE THEORY

The predictions about polymorphism with neutral alleles are derived from the infinite allele model outlined in section 7.5.2. The probability of fixation of a neutral mutation is the constant value u, the mutation rate. The time taken for fixation, however, increases with effective population size. Knowing these two quantities, it is possible to estimate the likelihood that two or more alleles will be present in the population at the same time. The number of different alleles possible is effectively infinite, so that given the right circumstances large numbers may coexist, and almost all individuals may be heterozygotes for different pairs of alleles. A quantity frequently measured is therefore the heterozygosity, represented by the letter H. The population will be more or less in Hardy–Weinberg equilibrium, and H is calculated as $1 - \Sigma q_i^2$ where q_i is the frequency of the ith allele, and q_i^2 therefore the frequency of the ith homozygote. Strictly speaking, this expression is only exact if N is infinite.

The next step in the derivation takes account of the degree of inbreeding likely to occur in a population of limited size N. Inbreeding is discussed in section 7.4.2. In a population of limited size, the probability that two alleles will be identical by descent, F, increases from generation to generation. The change is such that

$$F_n = \frac{1}{2N_e} + \left(1 - \frac{1}{2N_e}\right)F_{n-1}$$

Now, in the presence of mutation, an equilibrium will be reached if the tendency for F to increase is balanced by the tendency of alleles to mutate to some other form, thus reducing F. A fraction $(1 - u)^2$ of the population consists of pairs of alleles, neither of which have mutated. If mutation is counteracting the tendency for F to increase then F may achieve an equilibrium. At equilibrium we can write

$$\hat{F} = \left[\frac{1}{2N_e} + \left(1 - \frac{1}{2N_e}\right)\hat{F}\right](1-u)^2$$

Rearranging this equation and ignoring negligible quantities, we get, for the equilibrium level of inbreeding

$$\hat{F} = \frac{1}{4N_e u + 1}$$

This is the fraction of the population homozygous for alleles identical by descent. Consequently the equilibrium value of H, the fraction heterozygous, is $1 - \hat{F}$, i.e.

$$\hat{H} = \frac{4N_e u}{4N_e u + 1}$$

Heterozygosity may be predicted from the neutral allele assumptions if the effective population size and the mutation rate are known.

Given this model, it is not immediately obvious what the effective population size is. When the concept of N_e was discussed in section 7.4.1 it was associated with the size of a panmictic unit, usually a very small quantity compared with the number of living individuals in a species. If selection pressures may vary over short distances then this is the unit required in order to understand the factors affecting gene frequency. However, the present assumption is that selection is not experienced at all. Under these circumstances, the effect of migration becomes more influential and has to be reckoned with when estimating N_e. Migration, like mutation, has the effect of reducing the tendency of F to increase. In exactly the same way as in the argument above, we can find the equilibrium value of F by writing

$$\hat{F} = \left[\frac{1}{2N_e} + \left(1 - \frac{1}{2N_e}\right)\hat{F}\right](1-m)^2$$

where m is the migration rate. Consequently, taking account of the effect of migration

$$\hat{F} = \frac{1}{4N_e m + 1}$$

Mutation and migration have the same effect on the inbreeding coefficient. Migration rates have to be only as large as the mutation rate for it to be probable that every allele could be found in any part of the range. The effective population size to be used in neutral allele models is therefore based on the total extant population. As before, fluctuations in population from time to time, and in sex ratio and variance of offspring number, will affect the estimated number. We therefore have to consider population history, as well as current numbers, and N_e is the average total size of the species throughout its existence.

Now, u has been estimated from a wide variety of sources to be about 10^{-7} per year. It is therefore possible to estimate the expected amount of heterozygosity if we know the approximate value of N_e. By similar kinds of mathematical argument, the expected distribution of allele frequencies, the variance of H and the expected number of alleles in the population may also be estimated. These quantities may then be compared with the observed data.

Figure 12.1 shows the expected values of H for different values of N_e for

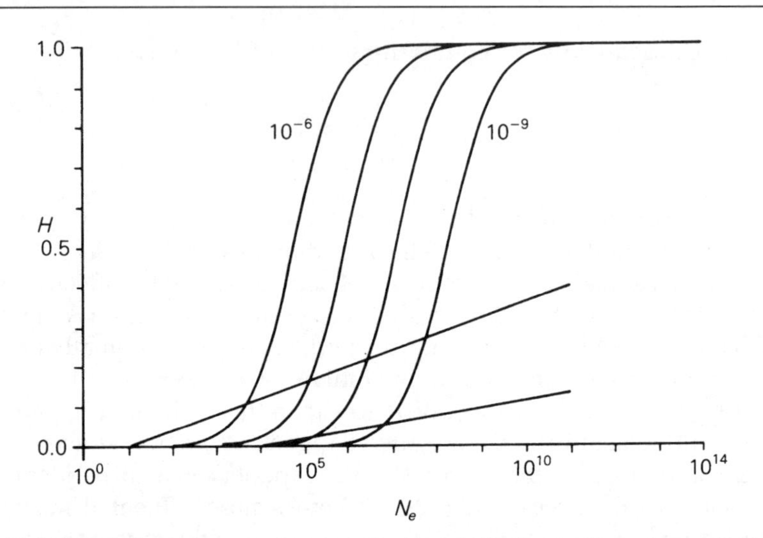

Fig. 12.1 The relation between heterozygosity, H, and effective total population size N_e. N_e is given in logarithms to base 10. The curves show the expected relationship on the neutral model for mutation rates of 10^{-6} to 10^{-9} per year. The straight lines enclose the area within which most observed values fall.

different estimates of the mutation rate u. The area between the two straight lines shows the probable position of the data points for samples from a range of species which have been compared with this expectation. The data are averaged over all loci, and the N values are guessed from all the evidence available. Even making allowance for the large errors which may be involved, it is evident that observation and expectation do not agree. Very abundant species have lower levels of heterozygosity, and rare species have higher levels, than predicted.

12.5 EXPLANATIONS FOR NON-AGREEMENT OF OBSERVED AND EXPECTED *H*

A straightforward explanation for the non-agreement is that the theory is wrong. If polymorphism was maintained by selection then there would be no necessary reason for heterozygosity to increase with population size and a fairly horizontal pattern might be expected. However, the neutral model cannot be dismissed as simply as that. In the first place, the effective population size depends on the number of individuals present through time in such a way that the estimate N_e is the harmonic mean of numbers in all generations (section 7.4.1). We know that the environment has not been constant over the duration of existence of species currently living, and it may be that species which are particularly abundant now have experienced bottle-necks in numbers in the past. Relatively rare species may once have been more abundant. If so, the N_e values attributed to the data should be contracted towards the centre of the distribution, and the disagreement is lessened.

Another consideration is that many mutations may be weakly deleterious in effect. If the disadvantageous effects were independent of environment then they would not affect the estimation of N_e. If the selective coefficients were not very different in size from $1/2N_e$, then the selection exerted becomes effectively density dependent. When populations are small, the selective coefficients would be smaller than the drift effects and such alleles would be neutral. In large populations, however, selective effects would be larger than drift effects, and would lead to elimination of the deleterious alleles. This would depress the expected curve of H on N_e.

12.6 CONCLUSION

Many other arguments have been ranged against the neutral theory, based on experimental data and logical deduction. Almost always it has been possible to put up a counter-argument from the neutralist position. Starting from examination of molecular variation, it seems reasonable to conclude that many alleles would be selectively equivalent in function. Many of the tests designed to demonstrate this have, however, failed to support or contradict this assumption. What may certainly be said is that the theoretical simplicity and predictiveness which the model first possessed has been lost. The neutral model requires as many *ad hoc* assumptions as selective ones.

12.7 SUMMARY

To explain genetic diversity the balance of mutation and drift is the alternative to balance involving selection. The non-selective or neutral theory has been applied most extensively to molecular variation. The mutation rate is assumed to be constant in time, with a rate of about 10^{-7} per year. The amount of polymorphism and the distribution and variance of allele frequency in a species can then be predicted if the total effective population size can be estimated. Divergence between taxa allows the time since they separated to be estimated, so providing a 'molecular clock' to measure the rate of evolution. It seems probable that the selection acting on molecular variation may often be low, so that these neutral processes operate. The agreement of data to model is not so good as to confirm this, however, and it is difficult even to arrive at a reasonable guess at the effective number in a species.

12.8 FURTHER READING

Crow, J.F. (1986) *Basic Concepts in Population, Quantitative and Evolutionary Genetics.*
Crow, J.F. and Kimura, M. (1970) *An Introduction to Population Genetics Theory.*
Kimura, M. (1983) *The Neutral Theory of Molecular Evolution.*
Li, W.-H. (1997) *Molecular Evolution.*
Nei, M. (1987) *Molecular Evolutionary Genetics.*

Gene interplay models for polymorphism 13

13.1 INTRODUCTION

One of the starting points of the neutral mutation theory is that selection moulds the phenotype while the genes are concerned with the mechanics of phenotype production. There is room for variation in frequency of genes without effective phenotypic change being involved. This loose causal linkage is also important in some other theories of polymorphism. The ideas discussed in this chapter have several origins, but they all involve the question of the nature of the interrelation of the genotype and phenotype.

13.2 NON-SEGREGATING CHARACTERS

We have seen that stabilizing selection, which removes the extremes from phenotypic distributions, is a very common phenomenon. It seems almost intuitively obvious that it should be so, because we would expect phenotypic characters to have optimal values, and variation from the optimum could occur in either direction to produce relatively less fit individuals. This does depend, however, on the nature of the character and on the origin of the variation. Some characters may be very closely related to fitness, such as the number of eggs produced or time taken to become an adult. In both cases, fitness could be linearly related to phenotype, so that it goes up progressively as fecundity increases or duration of development decreases. In these circumstances, selection will move the phenotype towards the optimal extreme value.

Stabilizing selection, favouring an intermediate optimum, is likely to be associated with characters less directly associated with fitness, or associated in a more complicated way. Increased body size in a mammal, for example, might be favoured because it brings with it increased fecundity. However, large size may be associated with changed metabolic balance which lowers the activity level, or it may simply ensure that the large animals are more subject to predation. The result of this balance of attributes is that intermediate values are favoured.

When fitness is linearly related to phenotype, we should expect that the heritability (that is, the selectable variation) of the character would be low, because disadvantageous alleles would have been eliminated. The genotype is homogeneous and consists mostly of dominant alleles for the characteristic conferring high fitness. Recessive alleles causing divergence of the character would tend to be eliminated as they segregated, minimizing the genotypic variance. We would not expect to find much evidence of polymorphism.

On the face of it, the situation should be different for stabilizing selection. Variants in the upper or lower direction should be relatively homozygous for increasing or decreasing alleles. These will be selected against, leaving individuals nearer the optimal phenotypic value, which are more likely to be heterozygous. This looks like a recipe for polymorphism by heterozygote advantage. In fact, however, an individual of intermediate phenotype may have a genotype consisting of multiple heterozygotes (*Aa, Bb, Cc, Dd* . . .) or one which is composed of equal numbers of homozygotes having opposite effects (*AA, bb, CC, dd* . . .). As a result of segregation, the first type produces offspring with a wider range of phenotypes than the second. On average, they have lower fitness because they are more divergent so that their genotypes tend to be eliminated in favour of the 'balanced homozygote' combinations. If the phenotype is controlled like this, then stabilizing selection will tend to reduce the genotypic variance and increase the average fitness of individuals in the population, which will tend to have the optimum phenotype.

The same result can be seen when there is dominance. Assume that the capital letters represent dominant alleles for increase in some measurable character such as a dimension, and lower case letters recessive decreasers. (This would be an example of reinforcing dominance; if some increasers and some decreasers were dominant then there would be opposing dominance.) Then if *A* contributes $+\frac{1}{2}$ and *a* contributes $-\frac{1}{2}$, and the same for other loci, two loci provide *aabb* $\rightarrow$ [– 2], *AABB* $\rightarrow$ [+ 2] and *AAbb* or *aaBB* $\rightarrow$ [0]. A cross of individuals of the two latter types, each with phenotypic value [0] would produce heterotic offspring with phenotypic value [+ 2]. An example of this type is provided by the work of Pooni and Jinks (1981) who crossed two inbred lines of *Nicotiana rustica*, measured for plant height. Although the F_1 showed heterosis ($86.6 \times 118.4 \rightarrow 140.3$), two extreme inbred lines selected from the resulting F_2 produced a new F_1 with a value lying between those of its parents ($75.9 \times 160.3 \rightarrow 138.4$). If an intermediate phenotype was optimal and there was inbreeding the system would tend to homozygosity.

Neither directional selection nor stabilizing selection maintain polymorphism on their own, yet the evidence from selection experiments and from observation of natural variation suggests that polymorphisms for alleles influencing quantitatively varying characters are common. The reason could be related to the way the genetic system operates. The alleles at the different loci involved are assumed to have equivalent functions. This is central to the theory of quantitative genetics. Alleles at different loci are therefore neutral with respect to one another, in the same way as alleles discussed in the previous section are neutral. If a character had been subject to stabilizing selection for a long time, most of the genotype would consist of 'balanced homozygotes' (i.e. *AA*, *bb*, etc., in equal proportions). This tendency would be opposed by mutation, and each mutation, when it occurred, would shift the phenotypic value away from the optimum, so that the mutations would be mildly deleterious. It is therefore possible that the polymorphism could involve mutation balanced by stabilizing selection.

The theory of such a balance has been examined by Kimura, among others. The method involves investigating the probability of mutation per locus, the amount of drift, the number of loci and the effect on fitness of the change in phenotypic value caused by allelic substitution. The number of polymorphic loci can be estimated by looking at the amount of divergence of selected lines in relation to the additive variance. Estimates for bristle number characters in *Drosophila melanogaster* produce figures of 50–100 loci, which is consistent with the idea that several different effects on the phenotype (e.g. body size, fecundity, metabolism, etc.) combine to determine the fitness distribution of a character. On the other hand, these estimates come from laboratory experiments which are designed to maximize the number of segregating loci, and representative figures from the field may be lower because more loci are monomorphic. Nevertheless, the picture emerges of a situation where allelic polymorphism and phenotypic variability result from a balance of drift, mutation and stabilizing selection on many loci with very slightly deleterious alleles.

13.3 OUTBREEDING AND DELETERIOUS MUTANTS

We have seen in Chapter 11 that in a diploid population, deleterious recessive mutants may be present at the relatively high equilibrium frequency of $\sqrt{u/s}$, where u is the mutation rate and s the selection coefficient of the mutant homozygote. The load has a value of u per locus, and since there are thousands of loci, the overall load is potentially appreciable. Inbreeding makes matters worse; each increase in the inbreeding coefficient F increases the mutational load. The observation of inbreeding depression, or general loss of viability, in animals and plants maintained in small populations is evidence that the effect may be severe. We could conclude, therefore, that any mechanism which increases outbreeding will be selected to minimize the depressing effect of deleterious mutants on viability. In the process, loci are maintained in a heterozygous state which would otherwise have become homozygous. The selection favouring outbreeding may be strong, but it would not be possible to ascribe it to the individual loci carrying the deleterious mutants. In terms of the population genetics of polymorphism, mechanisms favouring outbreeding serve to increase the effective population size and so increase the chance that other mechanisms will lead to polymorphism. If a gene is closely linked to the locus promoting outbreeding then it may actually be maintained in a polymorphic state for a long period.

13.4 GENETIC HOMEOSTASIS

When inbred lines are crossed and the resulting progeny show a higher fitness than the parental lines, they are said to exhibit heterosis. The reason may be the one described above; loci which were homozygous and relatively deleterious are rendered heterozygous. Each deleterious recessive is matched with a dominant of higher fitness and the fitness of the F_1 is raised above that of either parent stock although no locus need exhibit heterozygote advantage. This is spoken of as relational overdominance. Another reason for heterosis is that many loci actually possess heterozygote advantage. It could be that the resulting raised fitness is no more than the sum of heterozygote advantage ascribable to the individual loci. When quantitative geneticists have studied the effect, however, it has seemed more likely that heterosis is due to the covering of disadvantageous alleles than to heterozygote advantage *per se*.

There is, however, another theme that has been developed in this field. In the 1950s I.M. Lerner suggested that heterosis might develop because genotypes consisting of a large fraction of heterozygous loci presented a wider range of metabolic pathways than homozygous ones, so that development was better buffered against effects of environmental fluctuation. He termed this genetic homeostasis. The inference is that the heterozygous genotype should on average permit an individual to develop into an adult nearer the optimum phenotype than a more homozygous one. Since the buffering comes about as a result of the action of an ensemble of loci, genetic homeostasis would not necessarily be detectable by examining single loci. We would also have to look at epistasis between an array of loci, which is difficult to detect.

The idea of genetic homeostasis remained for some time an attractive one for which there was no good positive evidence. More recently, however, many groups have examined the possibility that a correlation of phenotypic value and genetic heterozygosity could be found by studying enzyme polymorphisms. Enzyme variability can be scored using electrophoresis, but what characters should we use to characterize the phenotypic variation?

Most animals are bilaterally symmetrical, so that it is a fair guess that the symmetrical form is near the optimum and asymmetry represents divergence from it. Development rate is another character which it is safe to assume bears a close relation to fitness. If organisms develop slowly they are likely to be less fit than rapid developers in the same population. A test of genetic homeostasis may therefore be made by comparing these measures of closeness to the phenotypic optimum with the amount of heterozygosity over a series of enzyme loci.

Many studies have shown that a correlation may exist. Michael Soulé examined asymmetry in populations of a lizard species. This is facilitated because lizards have convenient scale rows which can be counted on each side of the body. He then measured polymorphism, and showed that populations with the highest levels of asymmetry also had the lower average heterozygosity. Although the result supports the hypothesis, it is only a general association and does not demonstrate that the asymmetrical individuals were actually the relatively homozygous ones. This has been shown in many studies, however, for example in one on trout. Fish are also endowed with easily countable characters, and five of these (fin rays, gill rakers, etc.) were scored to measure left–right asymmetry. Each scored individual was then examined for heterozygosity. Forty enzyme loci were screened, of which 10 were polymorphic and five of these were scorable. The result is given in Table 13.1.

There is a clear correspondence between proportion of asymmetrical characters and degree of homozygosity; 2.25 characters are asymmetrical when none of the loci are heterozygous, and this figure drops steadily to zero for the two individuals which were heterozygous for all five loci. There is no reason to believe that the loci studied actually control symmetry in any direct way, but they may indicate the average levels of heterozygosity in the genome. If so, the implication is that a degree of linkage disequilibrium has built up in the genome.

Other studies on fish have failed to show the relation, but work on a

Table 13.1 Number of heterozygous loci and asymmetrical characters for 50 individuals in a population of rainbow trout

Asymmetrical characters (no.)	Heterozygous loci (no.)					
	0	1	2	3	4	5
0	0	0	0	2	1	2
1	2	2	4	4	1	0
2	1	3	10	6	2	0
3	0	3	2	3	0	0
4	0	0	1	0	0	0
5	1	0	0	0	0	0
Mean	2.25	2.13	2.00	1.67	1.25	0.00

Reprinted with permission from *Nature* (Leary *et al.*, 1983). Copyright (1983) Macmillan Magazines Limited. See also Zouras and Foltz (1987), Mitton (1998).

wide variety of organisms from pine trees to man does so. In some cases, loci which have large effects are ones which have specific effects on metabolism. Only time will tell whether or not the correspondence of homeostasis and enzyme heterozygosity becomes firmly established. If it does, it could sometimes be that the relation simply shows that outbred populations exhibit fewer deviant traits than inbred ones. However, it is possible that a genotype with many enzyme heterozygotes enables the phenotypic optimum to be approached with a higher degree of success than one where the enzymes are mostly homozygous. The model is akin to the quantitative genetic one, except that the loci concerned are not assumed to be functionally equivalent. Stabilizing selection could therefore maintain polymorphism without the tendency to move from the multiple heterozygote to the balanced homozygote situation.

13.5 CHROMOSOMAL REARRANGEMENT AND POLYMORPHISM

Restricted patterns of recombination can lead to the preservation of favoured combinations of non-allelic genes so that each combination can be regarded as making a united contribution to fitness. This point was raised in section 9.6, where it was argued that epistatic selection can lead to the development of super-genes. We now consider how the evolution of such combinations can be brought about. Such restrictions are of little consequence to habitual inbreeders which have gradually fixed advantageous combinations of non-allelic genes in the homozygous condition. They may be common in outbreeders if there is continuous long-term selection for particular non-allelic combinations. If present, they assume prime significance when a normally outbreeding species is suddenly forced to inbreed (through either a change in mating system or a reduction in population size), when there is a danger of haphazard disadvantageous combinations becoming fixed as homozygotes.

There are two ways in which recombination may be restricted at meiosis: one direct and the other indirect. Direct mechanisms result in fewer recombinants being produced whereas indirect mechanisms prevent recombinants from giving rise to viable gametes. Direct mechanisms rely on restricted chiasma frequency and/or distribution, either as a result of developmental control of chiasmata themselves or as the consequence of restricted pairing affinity. By contrast, indirect mechanisms ensure that recombinant chromatids are genetically unbalanced, often being duplicated for some segments while deficient for others.

13.5.1 CHIASMA LOCALIZATION

Many species have normal karyotypes and yet show extreme patterns of chiasma localization at meiosis. Characteristically, such localization is concentrated at the ends of the chromosomes so that chiasmata are predominantly distal in location but some species show the opposite pattern with the majority of chiasmata being proximal to the centromere. Probably the most dramatic and extreme example is provided by the Palmate newt *Triturus helveticus* which has metacentric chromosomes and no obvious sex chromosomes. Here, chiasmata are restricted to segments very close to the centromeres in females but very close to the ends of the chromosomes in males. The consequence of this arrangement is that most of the genes in each chromosome arm are inherited *en bloc*. Such localization of chiasmata is likely to have evolved in response to diminishing populations in contracting habitats, probably during the Pleistocene but perhaps aided by land drainage and other human activity. Beneficial combinations of genes are maintained and diversity reduced.

Localized chiasmata are more typically associated with the heterogametic sex in animals. Thus in the giant marsh grasshopper *Stethophyma grossum* chiasmata are proximally localized in heterogametic males but show no evidence of localization in homogametic females. Many plants show localized chiasmata but evidence is usually only available from studies of male meiosis, female meiosis being difficult to observe. Species such as rye *Secale cereale* have chiasmata in predominantly distal locations whereas the snake's head fritillary *Fritillaria meleagris* has chiasmata proximal to the centromere. In these species, localization of chiasmata seems to result from the pattern of pairing initiation, proximal segments being the first to pair in the fritillary but distal segments being the first to pair in rye.

13.5.2 INVERSION POLYMORPHISM

Beneficial combinations of non-allelic genes can be protected from recombination if they are included within inverted segments. The position of the inversion in relation to the centromere is of critical importance at meiosis. If the inversion is contained within a single chromosome arm, it is said to be paracentric (i.e. to one side of the centromere). If the centromere is included, the inversion is said to be pericentric. Crossing-over in homozygotes does not of course destroy the sequence, while if it occurs between the standard and inverted sequences, loss of fertility may ensue.

Crossing-over within the inverted segment of an inversion heterozygote will lead to duplication/deficiency for the recombinant chromatids. One recombinant strand will be duplicated for the segment to the 'left' of the inversion but deficient for that to the 'right' whereas the other recombinant strand will be deficient for the segment to the 'left' but duplicated for that to the 'right'. Neither strand will be genetically balanced and any gamete to which it passes will be inviable. The significance of the position of the centromere is that a chiasma within a heterozygous paracentric inversion will give rise to recombinant strands which show duplication/deficiency for the centromere. These structurally impaired chromatids will show delayed segregation – if they segregate at all.

The dicentric chromatid forms a bridge at the first anaphase whereas the acentric chromatid forms a fragment which lags behind at anaphase I (Figure 13.1(a)). Most animals and plants exhibit a type of female meiosis which results in four products arranged as a linear tetrad. Normally, only one product at the favoured end of the tetrad goes on to form the egg, the other three spontaneously abort. Heterozygosity for a paracentric inversion thus ensures that if any recombination does take place within the inverted segment, the products of that recombination do not reach the end of the linear tetrad and hence fail to reach the egg. No such mechanical considerations would prevent loss of male fertility, where the second meiotic division forms in a plane at right angles to that of the first division. In the majority of advanced Diptera, however, no chiasma forms during meiosis, so that no genetically unbalanced chromatids are formed.

Paracentric polymorphism is therefore characteristic of many dipteran species. Paracentric inversion polymorphisms have been studied most extensively in species of *Drosophila*. Some inversions are monomorphic and these serve as genetic isolation mechanisms in many of the species. Some species such as *D. virilis*, *D. simulans* and *D. repleta* show no inversion polymorphism, while others such as *D. robusta* and *D. willistoni* and *D. subobscura* show extensive polymorphism. In *D. willistoni*, for example, over 50 different inversion polymorphisms have been identified involving each of the three largest chromosomes with more or less equal frequency. In *D. pseudoobscura*, on the other hand, the majority of the 22 inversion polymorphisms known are found on chromosome III. The largest chromosomes in most species of *Drosophila* usually have near-terminal centromeres so that, where paracentric inversions are present, they often extend over substantial segments of the long chromosome arm. The inversions are named after the populations where they were first detected. They may be simple, as in the case of the arrowhead inversion in upland populations of *D. pseudoobscura* in the western USA compared with the standard lowland form, or complex as in the Chiricahua variant which is an inversion within an inversion. Seasonal and geographical variations in the frequencies of three types of arrangement are displayed in Figure 13.2. The selective basis of these fluctuations is considered in section 11.5.

Where inverted segments include the centromere, they often change the arm ratio of the chromosomes involved. Pericentric inversion heterozygotes are therefore often detectable at mitotic metaphase via the presence of asymmetrical (heteromorphic) pairs of homologues. At metaphase I of meiosis, these pairs form asymmetrical bivalents (Figure 13.1(b)). Heterozygosity for large pericentric inversions is very unlikely to give rise to polymorphism, since any chiasma forming within the inverted segment would lead to a 50% drop in fertility in females as well as males. Critically

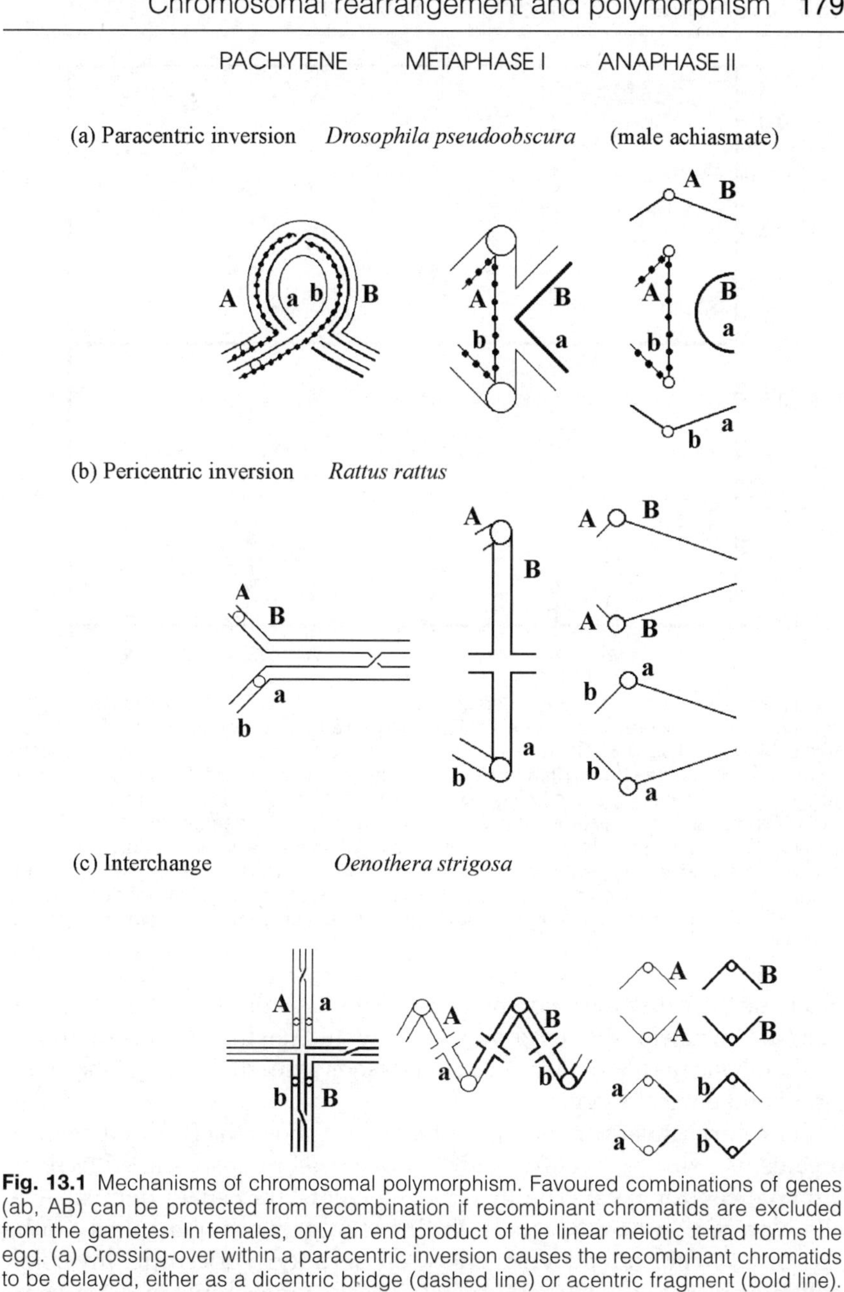

Fig. 13.1 Mechanisms of chromosomal polymorphism. Favoured combinations of genes (ab, AB) can be protected from recombination if recombinant chromatids are excluded from the gametes. In females, only an end product of the linear meiotic tetrad forms the egg. (a) Crossing-over within a paracentric inversion causes the recombinant chromatids to be delayed, either as a dicentric bridge (dashed line) or acentric fragment (bold line). (b) Only a special class of pericentric inversions can become polymorphic because crossing-over within the inverted segment would not lead to duplication/deficiency for the centromere. Small inverted segments around the centromeres of acrocentric chromosomes are unlikely to pair so that crossing-over does not take place there. The resulting bivalent is heteromorphic (asymmetrical) at metaphase I. (c) Only zig-zag (alternate) multivalents guarantee fertility in interchange heterozygotes; these ensure that random assortment does not take place between the standard and interchange chromosomes.

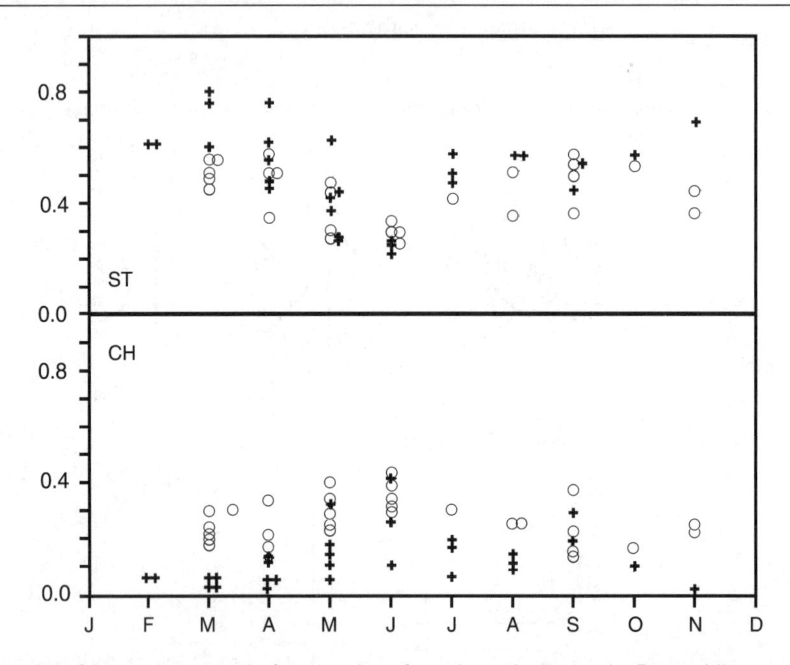

Fig. 13.2 Seasonal change in frequencies of two inversion types in *Drosophila pseudoobscura*. The data were collected by Th. Dobzhansky and his associates in the Piñon Flats area of Mount San Jacinto, California, in a long-term study. Records from 1939 to 1952 are represented by circles, from 1953 to 1970 by crosses. The standard (ST) inversion falls in frequency in summer, its place being taken by Chiricahua (CH), another inversion arrowhead (AR, not shown) which fluctuates in parallel with CH and some rarer forms. ST is a lowland form while the others are characteristic of higher altitudes. Furthermore, ST appears to show more extreme seasonal fluctuation in later years (crosses) than in earlier years (circles). These field observations led to the investigation of possible modes of selection and maintenance of polymorphism referred to in section 11.5. Data from Dobzhansky (1971).

in females, because the recombinant chromatids would not show duplication/deficiency for the centromere, they would not be structurally abnormal and hence as likely as the parental strands to reach the end of the linear tetrad and enter the egg.

Pericentric inversion polymorphisms are most likely to be seen in organisms with acrocentric chromosomes (chromosomes with near-terminal centromeres). The inverted segments are usually too small to allow pairing in heterozygotes. In this way, genic combinations can be preserved with no adverse effect on fertility. Pericentric inversion polymorphisms occur in orthopteran insects, for example in the Australian grasshopper *Caledia captiva* and the Californian grasshopper *Trimerotropis pseudofasciata*. A very similar pattern is exhibited by populations of the black rat *Rattus rattus*, many of which are subdivided into small inbreeding cohorts. Here the Asian race with $2n = 42$ chromosomes shows polymorphism for three tiny pericentric inversions on different chromosomes. The frequencies of the standard and inverted segments vary with latitude, not only on a continental scale but also within the Japanese mainland (Yosida *et al.*, 1971).

13.5.3 TRANSLOCATION POLYMORPHISM

Structural mutations at the centromere which result in two telocentric chromosomes fusing to form a single metacentric chromosome are known as Robertsonian fusions. Individuals which are heterozygous for Robertsonian fusions exhibit V-shaped trivalents during meiosis. Such trivalents are formed as a result of the pairing of each telocentric with the homologous arm of the metacentric and ensure genically balanced segregations in the fusion heterozygote. Robertsonian fusions reduce the number of linkage groups and so restrict recombination between erstwhile unlinked combinations of genes. Where such combinations show linkage disequilibrium, the fusions will confer a selective advantage and so rise in frequency. The resulting polymorphism may be transient, ultimately giving rise to a new monomorphic karyotype and a new cytological race (see Chapter 16), or it may be balanced, with the frequencies of the old and new arrangements varying in response to changes in habitat. Polymorphisms for Robertsonian fusions occur in the grasshopper *Oedaleonotus enigma*, the common shrew *Sorex araneus* and members of the *Tradescantia* family (Commelinaceae) such as *Gibasis scheidiana*.

Combinations of unlinked genes can be protected by reciprocal translocations of segments between non-homologous chromosomes. Such reciprocal translocations are known as interchanges. The break-point may occur at the centromere, so that whole arms are exchanged, but it is more usually within a chromosome arm. Interchange heterozygotes will only be fertile if the interchange chromosomes are included in one set of gametes and the standard chromosomes in the other. Segregation of the interchange and standard chromosomes is ensured if the partially homologous chromosomes form zigzag (alternate) multiples at first metaphase of meiosis. Alternate orientation is promoted by equidistant centromeres and distal chiasmata (Figure 13.1(c)). It is thus most likely in organisms with metacentric chromosomes. Alternately orientated multiples have the unique consequence of preventing recombination between the basic and the interchange chromosomes. As a result of such restriction, beneficial combinations of genes can be maintained across several linkage groups. The chromosomes pointing towards a common spindle pole constitute a balanced genome known as an interchange complex. The American cockroach *Periplaneta americana* normally occurs as large outbreeding populations, showing regular bivalent formation at meiosis. Small populations found in Welsh coal mines are, however, forced to inbreed (brother–sister mating, etc.) and here interchange multiples are commonly seen at meiosis.

The trend towards interchange polymorphism with decreasing population size is clearly illustrated in marginal populations of the Californian peony *Paeonia californica*, the western Australian desert plant *Isotoma petraea* and especially in the species of evening primrose (*Oenothera*) which are large-flowered, bivalent-forming outbreeders in western North America but mainly small-flowered, multivalent-forming inbreeders in the east. In some species, the interchange is so large that it includes all the chromosomes and every individual is an interchange heterozygote (e.g. *Oe. muricata*). Permanent interchange heterozygotes contain two complementary interchange complexes and pass equal numbers of each complex to their gametes. In theory, therefore, half the progeny on selfing should be homozygous for one complex or another and so form only bivalents at metaphase I. In practice, all the progeny are interchange heterozygotes. The homozygotes either die due to the presence of zygotic lethal genes, which are expressed in the sporophyte but not the gametophyte, or they are never formed because gametic lethals of one complex kill the male gametophyte whilst those of the other complex kill the female gametophyte. The complementary gametic lethals are said to be balanced.

Examples of each type of lethal system are provided by species of *Oenothera*. The interchange complexes in *Oenothera* have been given Latin names, according to their morphological effects in interspecific crosses. When the interchange heterozygote *Oe. lamarckiana* is crossed with the standard homozygote *Oe. hookeri*, half the progeny are attractive densely flowered plants with hairless leaves while the other half have more lax inflorescences and narrow hairy leaves. Both classes of progeny are produced in equal numbers, whether *Oe. lamarckiana* is the female parent in the cross or the male. Otto Renner, who first advanced the theory of genomic complexes to explain these unusual patterns of inheritance, called the complex responsible for the more attractive progeny *gaudens* (joyful) and the other *velans* (coated). While *Oe. muricata* also produces two types of progeny in interspecific crosses, one type with nodding stems is only produced when *Oe. muricata* acts as male parent in the cross and the other, with stiff stems, is only produced when it acts as female parent. Renner respectively named the complexes responsible *curvans* (nodding) and *rigens* (stiff). The segregation of these types in the backcrosses indicate that the alternative interchange heterozygotes are heterozygous at a number of loci within them.

Oenothera lamarckiana forms homozygotes of both *velans* and *gaudens* but both are inviable so that only 50% of fertilized eggs survive and all are interchange heterozygotes, bearing balanced zygotic lethals. *Oenothera muricata* only forms interchange heterozygotes. The *rigens* complex contains a gametic lethal which kills the pollen but not the embryo sac. This results in 50% of the pollen being inviable and the viable pollen all bearing the *curvans* complex. The situation on the female side is more complicated.

Normally, female meiosis results in a linear tetrad of potential megaspores but only that at the micropylar end of the tetrad forms a functional megaspore and develops into the embryo sac. It appears to be at a positional advantage. Because the two complexes segregate at the first division of meiosis, the megaspore at the micropylar end of the tetrad will bear a different interchange complex from that at the opposite (chalazal) end. If the megaspore at the micropylar end bears the *rigens* complex, it will develop normally into a viable embryo sac. If the *curvans* complex is at the micropylar end, it will begin to develop but so will the megaspore at the chalazal end which bears the *rigens* complex. Eventually, the *rigens* megaspore will out-compete and replace the *curvans* megaspore. This process, known as the Renner effect, was discovered by Renner in 1921. It has the important consequence that the *rigens* complex is not transmitted through the female gametophyte and yet there is no loss of female fertility. The situation in both species is summarized in Table 13.2.

Evidently, the system of balanced gametic lethals in *Oe. muricata* is less wasteful of ovules than the system of zygotic lethals in *Oe. lamarckiana*. In evolutionary terms, we can envisage how a system of gametic lethals could arise and replace a system of zygotic lethals but the reverse process does not seem possible. It is of interest in this connection that *Oe. lamarckiana* has not gone as far down the path of permanent interchange

Table 13.2 Comparison of zygotic and gametic lethal systems in two species of *Oenothera*, each a permanent interchange heterozygote. Columns represent pollen types and rows embryo sac types. *Oenothera lamarckiana* is an example of a zygotic lethal and *Oe. muricata* a balanced gametic lethal. In the latter, only one type of pollen and one type of embryo sac is viable

	Oe. lamarckiana (syn. *Oe. erythrosepala*)	
	Gaudens	*Velans*
Gaudens	*Lethal*	*Guadens velans* heterozygote
Velans	*Gaudens velans* heterozygote	Lethal
	Oe. muricata (syn. *Oe. parviflora*)	
	Curvans	*Rigens*
Curvans	–	–
Rigens	*Curvans rigens* heterozygote	–

heterozygosity as has *Oe. muricata* because only 12 of the 14 chromosomes are involved in interchange whereas all 14 are involved in *Oe. muricata*.

The best-known example of a permanent interchange heterozygote is *Rhoeo discolor*, a member of the Commelinaceae (the family including the genus *Tradescantia*) which is commonly grown as a house plant because of its attractive purple-backed leaves. This species has no close relatives and regularly self-pollinates, the pollen being released while the flower is still closed – a situation known as cleistogamy (from the Greek meaning closed reproduction). At zygotene, all 12 chromosomes pair to form a single multivalent. A ring of 12 is often seen at metaphase I, showing a distal chiasma in each of the 12 pairing segments. Smaller associations may be present if fewer than 12 chiasmata have formed. Despite its habitual selfing, *R. discolor* remains permanently heterozygous, not only for the interchanges but also for the chromosome segments on either side of the zig-zag chain. Such hybridity almost certainly persists through one of the two systems of lethal genes similar to those found in permanent interchange heterozygotes of *Oenothera*.

Many animals and plants have supernumerary chromosomes which are not present in all individuals and which are usually not associated with any gross phenotypic effects. Such chromosomes are known as B-chromosomes to distinguish them from the standard complement of A-chromosomes. They may or may not differ from the standard complement in size or structural organization; unlike the A-chromosomes, they are characterized only by their variability. B-chromosomes often have interesting geographical or ecological distributions which suggest that they are not selectively neutral. They are frequently found in central rather than marginal populations, as in the greater knapweed *Centaurea scabiosa* and the short-horned grasshopper *Myrmeleotettix maculatus*. The genetic and ecological significance of B-chromosomes is likely to depend on the genes which they carry and how these interact with the background genotype of the A-chromosomes. One general possibility is that B-chromosomes may disrupt the meiotic behaviour of A-chromosomes, especially by interfering with the control of chiasma formation. Such interference could conceivably have consequences similar to those of A-chromosome polymorphism but there is no direct evidence for such a possibility.

These varied examples show that beneficial combinations of genes can be protected from recombination by restrictions on chiasma formation or by chromosomal rearrangements. Such restrictions on recombination become especially important when normally outbreeding species are forced to inbreed. Mechanical considerations indicate that inversion polymorphism is more likely to become established in organisms with acrocentric chromosomes whereas interchange polymorphism is more likely to evolve from metacentric karyotypes. B-chromosome polymorphisms may reflect the action of selection on recombination but their general evolutionary significance remains an enigma.

What can be said in general, however, is that the mechanisms described generate a further level of genetic variability. Genic mutations are usually eliminated but are sometimes advantageous and result in substitutions or polymorphism. In a manner analogous to mutation, structural reorganization of chromosomes occasionally takes place during meiosis. The rearrangements are likewise usually eliminated, but sometimes the interplay of selection at the genic level favours the new arrangement and may result in structural polymorphism. Crossing-over may be reduced or eliminated so that different gene associations develop on the alternative chromosomes undisturbed by recombination. The new chromosomal organization promotes further genic heterozygosity.

13.6 SUMMARY

Continuously varying characters have a genetic variance and are subject to stabilizing selection. The variance indicates genetic variability, but the selection does not itself maintain it. It is probable that polymorphism for the alleles concerned is the result of a balance of mutation and drift, with selection eliminating the extreme combinations.

Almost all organisms have some sort of genetic recombination system, usually involving different sexes. Whatever the origin of these systems, genetic variability may be partly a consequence of their existence. Deleterious recessive mutants are likely to be heterozygous, so that they are less rapidly eliminated than they would be in an inbreeding species, and on average the outbreeding species has a higher fitness (since it exhibits relational overdominance). Having many heterozygous loci may also be an advantage in terms of gene action during development or in a fluctuating environment (genetic homeostasis). Experimental evidence of associations between enzyme polymorphism and high symmetry or low variance of phenotypic characters may be interpreted in either way.

Like genes, chromosomes may mutate to form new structural forms by inversion, breakage and interchange. Interplay between these and selection at the genic level may result in structural polymorphisms. Selection during the replication process sometimes results in structural heterozygotes becoming permanent. Structural heterozygotes represent a level of genetic diversity in their own right, and they also allow further genic diversity to accumulate.

13.7 FURTHER READING

John, B. (1976) *Population Cytogenetics.*
Mitton, J.B. (1998) *Selection in Natural Populations.*
Sheppard, P.M. (1975) *Natural Selection and Heredity.*

Polymorphism: a case history 14

14.1 CAN WE EXPLAIN SPECIFIC CASES?

Various organisms have made their appearance in previous sections to illustrate ways in which polymorphism may be maintained. We might equally ask whether the polymorphism in a selected species has been satisfactorily explained. With respect to Batesian mimetic butterflies such as *Papilio dardanus* or *P. memnon*, the answer is a reasonably confident yes. Their models are undoubtedly distasteful while the mimics are edible. The model patterns are conspicuous and the mimics employ a variety of devices to resemble them. Selective predation must give an advantage to rarity, so that establishment of a polymorphism is a natural consequence of adaptive coloration. Not all inferences of selection from association are as convincing, however. In South America there is a striking and conspicuous green and black butterfly, *Philaethria dido*, belonging to the distasteful family Heliconiidae. A green and black nymphalid species, *Victorina stelenes* which is probably edible, lives in the same regions and has a similar pattern. Pinned specimens look like model and mimic. Although their ranges overlap, however, these species do not occupy the same habitats and they do not resemble each other in their manner of flight. The nymphalid lives successfully on Caribbean islands where *Philaethria* is absent. There must therefore be some doubt as to whether selective predation led to this convergence.

A point about methodology arises from these studies. In scientific investigations we tend to give high value to experiment as a way of making discoveries. It is certainly possible to experiment with mimicry in butterflies. For example, captive or wild birds may be trained to avoid particular patterns by associating the image with a stimulus such as disagreeable food or, in some types of experiment, mild electric shocks. The difficulty with this approach is that, while positive, such experiments are likely to be over-simplistic. They may not allow generalization to natural conditions. To understand adaptations in nature, we often rely as much on inference from patterns and associations as on controlled experiment. This is also true, for example, of studies on industrial melanism or variation in shell colour in snails. A lot has been discovered about polymorphism in *Cepaea nemoralis*, but the pieces do not yet add up to a consensus view, which convincingly accounts for all the phenomena observed. Whereas it is easy to see that birds in experimental conditions can tell pink snails from yellow, it is not obvious why the snails should be so persistently variable.

14.2 THE *CEPAEA* PROBLEM

The shell colour and banding polymorphism in *Cepaea nemoralis* has been discussed in Chapter 9. The snails live in moderately sized, semi-isolated populations and occupy a wide range of types of habitat. One morph is better concealed than another in one habitat, while the relation is reversed in another. Observation of the action of predators and of change in morph frequency over time suggests that selective pressures are sometimes quite large at a few per cent per generation. With selection operating at that level, the polymorphism cannot result from balance of mutation and genetic drift, so that some kind of selective balance must be involved. A process which acts to maintain polymorphism, as distinct from merely changing frequency, is sometimes called a centripetal process. We have seen that frequency-dependent selection by predators may fall into that category as could heterozygote advantage, present for some reasons connected with non-visible attributes of the morphs. A balance could also be set up between different selection pressures in different places with migration between them.

When work on *Cepaea* was at its height, reviewers tended to prefer different candidates for the centripetal process. Clarke *et al.* (1978) favoured frequency-dependent predation. Cain (1983) considered heterozygote advantage to be the underlying factor, the existence of which allows the influence of other selective forces to become apparent. Goodhart (1987) argued that the chromosomes exhibit heterosis, or net heterozygote advantage, while concluding that the actual colour of an individual made little difference to its fitness. Jones *et al.* (1977) did not feel a clear choice could be made, but listed eight forces known to operate. The climate of the region in which a population is situated (1) may affect the average gene frequency, while (2) the local topography, e.g. exposed hillside versus valley bottom, also plays a part. Predators (3) selectively remove the most conspicuous morphs, so that the population becomes relatively cryptic. By frequency-dependent searching (4), predators tend to protect the polymorphism, while the way they do so is influenced by population density (5) and presence or absence of the congeneric species *C. hortensis* (6).

Migration from an adjacent area subject to different selection (7) will modify frequency, and historical accident may create a particular linkage disequilibrium (8), affecting the way the population responds to selection. Before the influence of these options can be assessed we certainly need to know more about three of them, heterosis, visual selection and the amount and effect of migration.

14.3 HETEROSIS

In slightly different ways, Cain (1983) and Goodhart (1987) argued that heterozygote advantage or heterosis supplies the necessary stable equilibrium. Although such arguments now get little airing, this is not because they have been shown to be untrue, rather that they are not confirmed as widespread. Systems such as the inversions in *Drosophila pseudoobscura* or the mimetic polymorphisms in papilionid butterflies, undoubtedly demonstrate the evolution of heterotic super-genes and thus illustrate an important class of evolutionary event. It is possible that similar systems exist in snails. A large number of species are polymorphic, whether for distinct visible characters or not. The evolutionary persistence of a heterotic system could be of very long duration, longer, in fact than the life of a species. This would explain the presence of shell pattern polymorphisms in many species of Helicidae, the family to which *Cepaea* belongs. The complement of heterosis is inbreeding depression. Although some snails, such as *Rumina decollata* (Subulinidae) or *Liguus fasciatus* (Bulimulidae) undergo extensive self-fertilization, helicids are almost exclusively outbreeders. A marked loss of fitness is known to be associated with inbreeding in two helicids which have been studied, *Arianta arbustorum* and *Helix aspersa*. The conventional way to measure heterosis would be to examine how variance in some fitness character changes with multi-locus heterozygosity, as indicated in section 13.4, but no experiments of the scale required have yet been made. In *Cepaea* shell size and shape affect fitness, shell dimensions are to some extent heritable and there is some evidence that they may be subject to stabilizing selection. An alternative approach is therefore to compare variance in individuals carrying dominant shell colour or pattern genes with those that are homozygous for the recessive, since the dominant group includes heterozygotes for the loci concerned and perhaps for sections of chromosome in which these loci are situated. Investigation of this type has shown a modest tendency for the dominant phenotype to have a lower variance of shell dimensions than the recessive one. Persistence of the visible polymorphisms is perhaps partly due to heterosis of chromosomes which include the colour and banding loci.

14.4 VISUAL SELECTION

Vertebrate predators appear to be especially good at detecting circular objects. For this reason, many prey species in unrelated animal groups have developed eyespot patterns which are used to confuse and deter predators. Snails may therefore be relatively poorly protected against predation because they are more or less round. Instead of being closely cryptic, the different morphs of *Cepaea* may present a set of images distinct from the background and from each other, which are difficult to interpret in a heterogeneous environment.

As indicated in Chapter 9, experiments show that predators often tend to select their prey in a frequency-dependent manner, overlooking whichever forms happen to be rare. General surveys support the impression that they do so when the prey are molluscs. Polymorphic pulmonate snails appear to be associated with the ecological, geographical and behavioural conditions listed in Table 14.1(a). The first five conditions suggest that polymorphic species are the ones on view and exposed to visually hunting predators. In European helicids and marine bivalves, apparent, exposed and colourful species are more likely to be polymorphic than hidden ones, which are unicoloured dark or white (Table 14.1(b)). Species in the prosobranch genus *Littoraria* tend to be confined either to the trunk and roots or the leaves of mangrove trees (Table 14.1(c)). These associations suggest strongly that polymorphism in molluscs is related to being visible against mixed backgrounds and exposure to predation by sight-hunting predators. Since the association is with polymorphism itself, rather than with brightness, the conditions evidently tend to protect polymorphisms. Polymorphism in *C. nemoralis* is therefore likely to matter because of the way the shells appear, and to be protected (to an as yet unresolved extent) by frequency-dependent predation.

14.5 POPULATION STRUCTURE

Colonies of *Cepaea* contain 100–1000 adults as a rule, sometimes as many as 20 000. Individuals have a strong inclination to stay put, only moving a few metres a year, so that each colony appears isolated from the next. This may not be so, however. Even small amounts of migration can greatly increase the effective population size, and long-distance dispersal may also have an important influence on polymorphism. Other types of observation suggest the snails are quite mobile. To get from somewhere south of the Pyrenees to the north of the continental range in Europe in the 6000–10 000 years available since the last ice age would require movement 10–100 times greater than that measured in experiments. Molluscs, in common with plants and other animal groups, colonized newly available territory with remarkable speed. We do not know how the movement was

Table 14.1 Some associations between polymorphism in pulmonate molluscs and various environmental, geographical and biological factors

(a) General associations, modified from Clarke *et al.* (1978)

Pulmonate molluscs are more likely to be polymorphic for shell colour and pattern:
1. in species which periodically or permanently climb above the ground than in those living on it,
2. in species of open habitats than in those of dense woodland,
3. if major predators hunt by sight rather than by smell,
4. in species of intermediate size than in very small or large species,
5. if the average density is high rather than low,
6. if the species outbreeds rather than inbreeds,
7. in warmer, wetter environments than colder, dryer ones,
8. in species of archipelagos than those of continental land masses,
9. in terrestrial than aquatic habitats.

(b) European Helicidae grouped according to favoured behaviour or habitat and general coloration of shell. Last row shows number polymorphic among those listed above it. Reproduced with the permission of National Museum of Wales from *J. Conch.*, Cain (1977), which gives details of the categories used

	Nocturnal or shady	More or less nocturnal	General habitat	Exposed	Very exposed
Dark	9	5	0	0	0
Medium	8	10	7	4	0
Light	0	1	2	10	17
White	0	0	0	1	3
Polymorphic	0	0	8	10	14

(c) Marine bivalves according to probable degree of visibility to predators and degree of colour and patterning. From Cain (1988)

	Well exposed to predators	Partially hidden	Hidden
Highly coloured and variable	10	11	0
Well coloured	12	20	1
Poorly coloured	7	26	12
White	4	17	44

(d) Association of polymorphism with substrate in species of *Littoraria* living on different parts of mangrove trees in the Indo-Pacific region. Data derived from Reid (1986)

	Trunk	Leaf
Monomorphic and variable	6	3
Polymorphic	0	8

achieved, but passive displacement is much more common in snails than the careful ecological work suggests.

Another way to assess the amount of migration is to look at data on variation in isozyme frequency between populations. A number of surveys have now been carried out on *Cepaea* which use measures of variance of allele frequencies at enzyme loci to characterize the degree of genetic differentiation between populations (e.g. Ochman *et al.* (1987), Selander and Ochman (1983)). Although some allozymes may be subject to selection, most alleles are probably more or less neutral or at least uncorrelated.

There is no reason to believe that the comparatively high correlation between samples is due to selection over all the enzymes examined. At any rate, these studies allow the variance in gene frequency between populations to be calculated.

In the enzyme data there are several alleles at each locus, but the results may be treated like a two-allele system if one allele is compared with the rest and change thought of in terms of forward and reverse mutation to and from the chosen allele. For a two-allele system with an allele at frequency q, inbreeding (F) is related to variance (V) as $F = V/[q(1-q)]$. In practice, q is estimated as the mean for the samples in a defined area or region in the generation examined. As time passes, F tends to increase. If the populations were completely isolated from each other, then in due course, balance of mutation and drift would result in a stable frequency distribution at which $Nu = (1-F)/8F$. This is the equation for equilibrium value of F in section 12.4, except that the number in the denominator is 8 rather than 4 because forward and back mutation at rate u is assumed, rather than an infinite number of alleles. The equilibrium is approached very slowly at a rate determined by the mutation rate.

Wright's island model for migration assumes that there are many populations, and that migration from any one to any other distributes individuals between them. On that basis, $Nm = (1-F)/4F$ at equilibrium (see section 12.4). When m is much greater than u it is therefore migration rate, rather than mutation, which determines the structure of the system. If there is less than about one migrant per generation, the populations behave as if they are distinct and tend to become homozygous. If there are more, however, populations tend to be heterozygous and share genes. J.F. Crow and others have shown that the inbreeding coefficient then moves quite rapidly towards equilibrium, taking a time of the order of hundreds rather than thousands of generations. In a field study of linked populations, variance between populations may not be too far from the equilibrium variance, so that estimates may be made from the equilibrium equation. Although population size is not known we can get some idea of Nm, the absolute number of individuals moving from one population to another. N is the effective population size, so the estimate is a minimum.

Table 14.2 shows estimates of Nm derived from survey results. They are fairly consistent, indicating movement of one or two individuals between populations per generation. If the populations are assumed to be completely isolated and long established, the quantity labelled m would be estimating the mutation rate, and effective population sizes would be two or three orders of magnitude larger than colony size indicated by ecological studies.

An alternative explanation of the comparative homogeneity of the colonies in a region is that, although isolated, they are recent and have had little time to diverge. A few calculations indicate that this assumption would

Table 14.2 Number of migrants, *Nm*, moving between colonies per generation, estimated from surveys of enzyme polymorphisms in *Cepaea nemoralis*. References to original work in Cook (1998)

Region	*Nm*	Data source
S. England, within areas	1.9	Johnson, 1976
England & Wales	0.2	Johnson, 1979
North Wales	1.1	Jones *et al.*, 1980
Pyrenees	2.1	Jones *et al.*, 1980
Within region, Pyrenees	2.4	Caugant *et al.*, 1982
Within valleys, Pyrenees	1.2	Ochman *et al.*, 1983
Within regions, Britain & Europe	1.1	Ochman *et al.*, 1987
French Pyrenees, populations within zones	2.6	Valdez *et al.*, 1988
Within regions, N. Spain	1.9	Vicario *et al.*, 1989
Within region, Pyrenees	0.3	Guiller and Madec, 1991
France & Iberia, within regions	1.2	Guiller and Madec, 1993

imply that regions of between 1000 and 10 000 square kilometres have been colonized in a matter of centuries. So again, extensive movement would have to occur. There is therefore either a high effective population number or rapid dispersal after frequent reductions in numbers or high average levels of movement after numbers are stabilized. Whichever explanation, or combination of them, most nearly describes the true picture, movement plays a greater part in the evolution of morph frequency patterns than indicated by ecological studies alone.

14.6 TIME SCALE

The habitats of *C. nemoralis* form a mosaic consisting broadly of woodland facies or open scrub and grassland. Almost all have been subject to massive alteration as a result of changes in land use and agriculture taking decades or centuries. In addition, climatic fluctuations over centuries or millennia have affected the species directly and through their effect on man. The study of fossils suggests that even in prehistoric times removal of trees, which for a snail mimics climatic change, helped *Cepaea* to disperse and become part of the northern fauna. In Britain, both *C. nemoralis* and *C. hortensis* first occurred in the Pliocene, disappearing when conditions became cold, to be followed by later reinvasions. They were present by 9000 years before the present, and abundant at the time of the climatic optimum about 6000 years ago.

The generation time is between 2 and 3 years, so that even under reasonably strong selection a large change in frequency will take decades. The duration of a particular habitat pattern is therefore similar to the time required for noticeable change in gene frequency. Changes in habitat occur at random with respect to the snails so that there is a continuous lag in response.

14.7 A POPULATION NETWORK MODEL

These considerations suggest that we should think of *C. nemoralis* as living as a network of populations linked by migration in a territory it has occupied for no more than 4000 generations. Migration between populations is much higher than direct ecological measurements suggest, as a result of passive displacement. Selective pressures, including selective predation, vary widely from one habitat to another, are often strong, and fluctuate periodically with time. Under these conditions, a network of a few thousand linked populations, each of a few hundred individuals, tends to move towards a flat distribution of morph frequencies, with few monomorphic populations, even when the starting frequency of one of the alleles is low. The reason is that accidental differences in selection cause divergences in frequency, while migration counteracts the tendency to diverge (Cook, 1998). Reasonable assumptions about selection and the amount of movement keep almost all populations polymorphic over the time scale concerned. There is no true balance, and such a system would ultimately become monomorphic unless mutation injected new variability, but the decay process is extremely slow when the number of populations is large.

14.8 CONCLUSION

Putting these different considerations together, we may now formulate an explanation for the observed pattern. First, some centripetal processes probably operate. There are circumstantial reasons to believe that both heterosis and frequency-dependent predation help to maintain polymorphism for shell colour and pattern. Evidence from a wide range of molluscs indicates that visible polymorphism is associated with apparency and visual predation, and numerous studies have shown that predators feed in such a way as to protect rare morphs in prey. Many helicids are polymorphic. They are outbreeders which sometimes, at least, suffer inbreeding depression. Chromosomal studies indicate the existence of localized chiasmata. These are good conditions for the visible characters to become linked into a balanced chromosome system. Both heterosis and selection by predators could be very ancient, acting over periods much longer than the lifetime of a single species. Response to the two types of selection could act synergistically on the genome to produce adapted super-genes, but the time scale involved must be very long and the selection is likely to be weak.

The second possible factor is that once polymorphism is attained it may persist for a very long time because selection in different directions in different populations interacts with migration between them. It is a truism to say that a species only has a temporary existence, or that a given territory is only temporarily available to it. If the time scale for loss of the polymorphism is similar to the time which the species has existed in a region, then there may be effectively permanent polymorphism although there is no true balance.

Cepaea species may therefore belong to a group in which evolution relating to breeding system and visual predation, operating for millions of years, have resulted in balanced heterotic genomes. Many types of habitat occupied are likely to select visible differences in shell colour and pattern. Over the shorter time scale of thousands of years these give rise to polymorphisms which persist but are not truly balanced. When persistent polymorphic states recur, they provide the genetic variability which allows a balanced system to develop.

14.9 SUMMARY

Cepaea nemoralis is polymorphic for genetically controlled shell colour and banding. Random drift, selective predation and climatic selection, both at a macro- and micro-scale, all affect gene frequency. The usual approach to understanding maintenance of the polymorphism has been to look for centripetal effects on frequency. Possible processes include balance of mutation pressure and drift, heterozygote advantage, relational balance heterosis, frequency-dependent predation, or some combination of these.

Mutational balance is overlaid by more substantial forces. There is some evidence for heterosis. Predation by birds probably protects the polymorphism. Although not substantiated for *Cepaea*, many studies show that predators behave in the appropriate manner, while shell colour polymorphisms in molluscs occur most commonly in species exposed to visually searching predators.

Migration between colonies is probably greater than originally thought. The present geographical range has been occupied for less than 5000 generations. Climatic and human modification alter snail habitats relatively rapidly, which in turn changes selection pressures. If thousands of populations are involved, migration coupled with selection which fluctuates but is not centripetal, may retain polymorphism for sufficiently long to account for the patterns we see today.

There may therefore be a two-stage basis to the polymorphism, comprising long-term but weak balancing forces coupled with fluctuating selection which does not necessarily balance but results in very slow elimination. Persistence of genetic variants in this way may provide the conditions for evolution of a balanced genome.

14.10 FURTHER READING

Clarke, B., Arthur, W., Horsley, D.T. and Parkin, D.T. (1978) Genetic variation and natural selection in pulmonate snails. In Fretter, V. and Peake, J. (ed.) *The Pulmonates, Vol. 2A, Systematics, Evolution and Ecology.*

Cook, L.M. (1998) A two-stage model for *Cepaea* polymorphism. *Phil. Trans. R. Soc. Lond. B*, **353**, 1577–1593.

Hanski, I.A. and Gilpin, M.E. (eds.) (1997) *Metapopulation Biology. Ecology, Genetics and Evolution.*

Jones, J.S., Leith, B.H. and Rawlings, P. (1977) Polymorphism in *Cepaea*: a problem with too many solutions? *Ann. Rev. Ecol. Syst.*, **8**, 109–143.

Genetic variability – conclusions 15

15.1 GENETIC CONTROL OF MUTATION

In taking stock of the multiple causes of genetic variability, it is as well to start from the beginning. Mutations arise from misreplication and repair failure, so that with the passage of time variation is injected into the lineages of all living organisms. One of the difficulties in determining the rate of input lies with definition. The smallest inherited change is the exchange of one base for another in a sequence of DNA. Estimation of the mutation rate in most eukaryotes has been achieved by recording detectable phenotypic changes determined by genes which may be composed of many hundreds of nucleotides, some of which influence the phenotype while some do not. Mutations may be caused by factors external to the organism, such as ionizing radiation. If all stages of the cell cycle are sensitive, the mutation rate will be constant in real time but proportionally higher per generation in organisms with long generation times than in those with short ones (this point is important in molecular clock calculations). Mutations occur at a higher rate in mitochondrial genes than in nuclear genes in the same organism because mitochondrial repair mechanisms are less effective. Chromosome reorganizations such as deletions, insertions or translocation are also included under the term mutation. Standard strains have long been available in *Drosophila melanogaster* which will detect both types of effect. For radiation-induced changes, the estimated rate of point mutations increases with dose roughly as the square root of the rate of induction of translocations, suggesting breakage as the mechanism, with translocation requiring two coincident breaks. Chemical mutagens do not show this relationship, however, indicating a different mode of operation. It is not surprising that there are great differences in rates measured for different organisms and systems.

Spontaneous mutation rates also vary widely between similar organisms, however, and between genes in the same organism. Rates as low as

4×10^{-10} mutations per cell per generation have been recorded for streptomycin resistance in *Escherichia coli*, whereas resistance to penicillin arises with a frequency of 1×10^{-7} in *Diplococcus pneumoniae*. Such low frequencies are only measurable in prokaryotes. In *Zea mays*, while no spontaneous mutations to waxy have been recorded, measured rates of mutation range from 1.2×10^{-6} mutants per gamete per generation for shrunken endosperm to 106×10^{-6} for aleurone colour. In *Drosophila melanogaster*, yellow body arises six times as frequently as ebony body (12×10^{-5} versus 2×10^{-5}). In mammals, estimates for the mouse are about 10^{-5} for coat colour genes and 10^{-7} to 10^{-6} for other loci. Figures for man are somewhat higher at 10^{-6} to 10^{-5} per generation for genetic diseases. Average rates of mutation of single bases assumed in theoretical studies of mutation, drift dynamics are about 10^{-7} per year.

Genetic variation resulting from mutation is only regularly apparent in prokaryotes. New mutants in prokaryotes are: (1) almost certain to arise in the huge populations, (2) bound to be expressed in the haploid condition and (3), rapidly propagated if advantageous. In *E. coli*, for example, a new source of streptomycin resistance is likely to arise at each generation in 2 μl of culture, since it contains at least 2×10^9 bacteria. If the antibiotic is present in the medium, the resistant mutants will be at a strong selective advantage. With a doubling time of 20 minutes in log phase at normal growing temperatures, more than four billion resistant bacteria could be produced within 11 hours (32 cycles of binary fission). Even unicellular haploid eukaryotes could not begin to match this flexibility in response to favourable mutation.

Mutation rates are known to be subject to genetic control and mutator genes have been identified in a wide range of prokaryotes and eukaryotes alike. A particularly impressive example is provided by the mut D5 mutator gene in *E. coli* (Table 15.1). This enhances the rate of reversion to wild type of deficient enzyme mutants by as much as 62 000 times that when its non-mutator allele is present. Moreover, every class of mutation is affected: base substitution (transitions and transversions), frameshift and suppression (in the tRNA genes). The gene is effectively functioning to effect repair after a deleterious mutation has occurred.

In prokaryotes reproducing asexually there will be strong selection for such repair mechanisms in a lineage carrying the defective target gene, while a lineage carrying the fully functioning gene would be under selection to minimize mutation rate. A condition for selection of the mutator effect is therefore that no reassortment of alleles occurs between the two loci. This requires either that the loci are side-by-side or that there is no recombination. It would be possible for a mutator locus to affect more than one target locus, so that the result of such selection could be to raise the average mutation rate. Another consideration also suggests adjustment of mutation rate. In asexual lineages, progressive improvement due

Table 15.1 Enhancement of tryptophan synthetase A gene (trp A) reversions to Trp^+ and of tRNA suppressor mutations by a mutator gene (mut D5) in *Escherichia coli*. pur, purine; pyr, pyrimidine. Abridged from Fowler *et al.* (1974)

Mechanism of reversion	Mutation	Mutator allele	Mutation frequency	mut D5/mut$^+$ mutation ratio
Transition	trp A446	+	3.1×10^{-10}	
G:C → A:T	↓			62 662
pur → pur	Trp^+	D5	1.9×10^{-5}	
pyr → pyr				
Homologous transversion	trp A11	+	8.6×10^{-10}	
G:C → C:G	↓			3325
pur → hom pyr	Trp^+	D5	2.9×10^{-6}	
pyr → hom pur				
Heterologous transversion	trp A88	+	6.1×10^{-10}	
A:T ⇌ C:G	↓			72
pur → non-hom pyr	Trp^+	D5	4.4×10^{-8}	
pyr → non-hom pur				
Frameshift	trp A540	+	2.3×10^{-8}	
	↓			170
addition of single base	Trp^+	D5	3.9×10^{-6}	
Suppression		+	1.7×10^{-10}	
	in a tRNA			424
trp A88 → pseudo	gene			
wild type		D5	7.6×10^{-8}	

to selection of advantageous mutants can only proceed to its fullest extent if two or more useful mutations at different loci occur in the same lineage. It thus depends on very improbable events, while genetic recombination can immensely speed up the process. This is an argument in favour of recombination, but where recombination is absent competition between lineages may act to modify mutation rates themselves.

Because of the enormous size of prokaryote populations, beneficial mutations are often present in sufficient numbers to respond rapidly to selection. Not only do most prokaryotes manage perfectly well without recombination, but recombination can be a bar to the control of mutation. By contrast, new genotypes are produced rapidly in eukaryotes by recombination, which shuffles the genic differences caused by mutation. In eukaryotes, it is not the rate of mutation which is normally controlled but that of recombination. The mutation rate may therefore be optimal in prokaryotes but it is minimal in eukaryotes.

15.2 GENETIC CONTROL OF RECOMBINATION

Most eukaryotic organisms rely on sexual reproduction and those that do not have a limited long-term future. Conjugation in some bacteria, such as

E. coli, is a sexual process but it is not full sexual reproduction in three important respects. First, it is an uncertain process, relying on the bacteria reaching a state of competence; we have no idea how frequent it is, either amongst species or natural populations of those species which have been observed to conjugate in the laboratory. Second, it is usually incomplete; the chromosome from the donor strain is rarely transferred in its entirety to the recipient. Third, while a limited amount of recombination takes place after transfer, there is no meiosis and, with only one main ring chromosome, no independent assortment. Evidently, there is a clear divide between prokaryotes and eukaryotes in the way genetic variation develops.

When most biologists think of sex, they think of mating and fertilization. If they think of recombination, they picture meiosis and particularly crossing-over. In fact, these two phases of the sexual cycle are part of the same process, and both equally important. Unlike binary fission, sexual reproduction produces a new generation which differs from the old because recombination has taken place. As we have seen in Chapter 13, sexual organisms can broadly be categorized into two types of breeding system. Inbreeding species have evolved balanced combinations of non-allelic genes, along with high levels of homozygosity. Outbreeding species, on the other hand, are highly heterozygous. When a normally outbreeding species is forced to inbreed, perhaps as a result of contracting numbers of individuals, the offspring show reduced vigour. This condition, known as inbreeding depression, is restored on subsequent outcrossing. These changes may occur (1) as a direct consequence of heterozygosity, (2) as a result of association and dispersion amongst polygenic systems or (3) as a product of non-allelic interaction.

Both breeding behaviour and meiosis are controlled by genes. The genetic control of breeding behaviour leads to restriction of inbreeding. In animals, this is mainly achieved via sexual dimorphism controlled by sex-determining genes on sex chromosomes. In fungi and flowering plants, it is usually due to self-incompatibility controlled by mating-type genes. Both systems will be reviewed in Chapter 18.

Every feature of chromosome behaviour at meiosis is under genetic control: synapsis (pairing), chiasma formation (crossing-over) and centromeric co-orientation (segregation). We have already seen some evidence in the Palmate newt, where chiasmata are proximal to the centromere in females but distal in males, and in the giant marsh grasshopper where they are proximal in the heterogametic male but unrestricted in the homogametic female (Chapter 13). Similarly, genetic control is evident where one sex fails to form chiasmata, as in many male dipterans including *Drosophila*, but the other is chiasmate. Interestingly, where chiasmata are restricted in one sex but not the other, that sex is always the heterogametic sex. Thus in Lepidoptera such as the silkworm

Bombyx mori, where the females are heterogametic and the male homogametic, it is the females which are achiasmate.

Numerous genes have been identified which control chiasma formation in a wide range of organisms where chiasmata are not normally restricted, for example in maize. Recessive mutants preventing chiasma formation in specific chromosomes have been identified in *Crepis capillaris* and *Hypochoeris radicata*, both plants in the dandelion section of the Asteraceae (Parker *et al.*, 1976). In most organisms, numbers of chiasmata (chiasma frequency) show quasi-continuous variation between cells. Such variation is likely to be subject to polygenic control as demonstrated, for example, in inbred lines of rye. It will also respond to selection. Shaw (1974) was able to select for low and high chiasma frequency in the desert locust *Schistocerca gregaria*. Each member of the chromosome complement was affected, with the exception of the smallest chromosomes in the low selection line. These normally only form a single chiasma, the minimum needed to ensure that homologous chromosomes are held together at first metaphase and so segregate regularly. The heritability was high, with about 50% of the variation in chiasma frequency being due to additive genetic effects. These genes are not subject to direct selection but their fate depends on the recombinants which they produce. In that respect, they resemble mutator genes.

15.3 GENETIC SYSTEMS

Given that genes control every aspect of recombination in the breeding system and meiotic system, we may ask how well the two systems are co-ordinated? We can examine this question by comparing the patterns of variation in chiasma frequency, exhibited by closely related inbreeding and outbreeding species. Inbreeders and outbreeders parcel their variation up in different ways. Outbreeding populations resemble F_2 progenies, most of their variation being within populations and much less between them. Populations of inbreeders resemble sets of inbred lines, most of the variation is between populations and little within. Over-zealous taxonomists sometimes split inbreeders into numerous 'species', each characteristic of a particular inbred line. Such groupings may well not interbreed under normal circumstances but they are not genetically isolated from one another, so that they are not true species. Taxonomy can sometimes confuse attempts to examine real biological patterns.

Zarchi *et al.* (1972) made a comparative study of patterns of variation in two wild diploid species of wheat: *Triticum longissimum*, a self-compatible inbreeder, and *T. speltoides*, a self-incompatible outbreeder. The species occupy similar habitats on disturbed ground and their distributions overlap extensively in Israel, where they were sampled. The patterns of chiasma variation matched those of morphological variation and were in line with expectation. Thus, *T. longissimum* exhibited highly significant differences between populations for close on a hundred morphological characters, whereas *T. speltoides* exhibited a less consistent pattern, some populations being significantly different and others not. As far as chiasma frequency was concerned, the inbreeder differed significantly between populations but not within them. The outbreeder showed no significant difference between populations but highly significant differences between half-sib families within populations. The data on these diploid wheats confirm that the genetic control of recombination at meiosis (via chiasma frequency) is integrated with that at fertilization. The genetic control of recombination throughout the sexual cycle is known as the genetic system.

One intriguing aspect of the data on the diploid wheats is the significant difference in mean chiasma frequency between the two species. The inbreeder *T. longissimum* forms an average of two chiasmata per bivalent, whereas the outbreeder *T. speltoides* forms only one and a half chiasmata per bivalent. This difference is consistent with others noted in a wide range of plant species, the inbreeders always having higher mean chiasma frequencies than those of the related outbreeders. On first consideration, this difference appears counter-intuitive. If inbreeding is about conserving favourable combinations of genes, why do inbreeders have more chiasmata than outbreeders? Since the genotypes are highly homozygous they will rarely produce recombinants, so that at least they are not obviously disadvantageous. High chiasma frequencies could be beneficial in the event of occasional outcrossing, producing a flush of novel recombinant genotypes from which new more vigorous lines may be selected. They are therefore at worst not disadvantageous and at best occasionally beneficial, but this does not explain how inbreeders arrive at high chiasma frequencies in the first place. The answer is provided by a comparative study of established outbreeding populations and their short-lived descendants.

Rees and Dale (1974) made a detailed study of variation in morphology, longevity and chiasma frequency in three species of forage grass: *Festuca pratensis* (meadow grass), *Lolium multiflorum* (Italian ryegrass) and *L. perenne* (perennial ryegrass). Their survey began with long-established populations around Aberystwyth in mid-Wales and extended along a transect northwards towards Liverpool. Longevity in grasses is inversely related to the proportion of plants flowering in their first year of growth from seed. Rees and Dale found that this proportion increased along their transect, as did the frequency of self-pollination. The trend towards populations increasingly consisting of inbreeding, short-lived plants, ran parallel to habitats which showed ever greater degrees of disturbance. As the populations became more inbred, so the phenotypic variance within populations declined, in chiasma frequency as well as in morphology. These observations were in line with those on the diploid wheats. Significantly, the mean chiasma frequency increased in line with progressive inbreeding and in opposition to the declines in phenotypic variance.

Evidently, the higher chiasma frequencies associated with inbreeding arise during the transition from outbreeding. When perennial outbreeding grasses are subjected to disturbed conditions which favour survival as seed, rather than underground stems, new genotypes which promote annual life cycles, selfing and hence high seed-set, will be at a strong selective advantage. Such genotypes will not be common at first because the parental populations consist of long-lived self-incompatible perennials. They will be most likely to be found amongst the progeny of highly heterozygous parents with lots of chiasmata. Because chiasma frequency is under genetic control, the genes for high chiasma frequency will be favoured during the transition to inbreeding. Ultimately, highly homozygous inbreeding populations will consist of genotypes fixed, not only for all the outward characteristics of inbreeding, but also for the high chiasma frequency. This hypothesis, advocated by Rees and Dale, provides the most plausible explanation for the high chiasma frequencies characteristic of inbreeding species. Once again, we see that selection does not act directly on the genes controlling chiasma formation but rather on the recombinants produced by such chiasmata.

We have already seen in Chapter 13 how structural changes in chromosomes can restrict recombination. In Chapter 16, we will investigate the role of numerical changes, in the form of polyploidy, in genetic isolation and speciation. Polyploidy arises when the chromosomes divide but the nucleus does not, a phenomenon known as restitution. Successive doubling of a diploid cell can give rise to a tetraploid and an octaploid. Crosses between diploids and tetraploids produce triploids; crosses between tetraploids and octaploids result in hexaploids. If all these events arise from within a single diploid species, they result in a series of autopolyploids with multiples of each chromosome type rather than pairs. The genotypes are correspondingly complex. Thus, given a diallelic locus *Aa*, there are five possible genotypes in an autotetraploid: nulliplex (*aaaa*), simplex (*Aaaa*), duplex (*AAaa*), triplex (*AAAa*) and quadriplex (*AAAA*). Notice that three of the five are heterozygotes. The polysomic nature of these genotypes leads to a complex pattern of heredity known as polysomic inheritance. An important consequence of polysomic inheritance is its resistance to inbreeding. The rate of loss of heterozygosity on selfing an autotetraploid, for example, compares with that produced by brother–sister mating in a diploid. Similarly, the loss of heterozygosity on selfing an autohexaploid occurs at a similar rate to that resulting from first cousin mating of diploids (Figure 15.1). The very complexity of polysomic inheritance impedes segregation so that autotetraploids may be said to be buffered against inbreeding.

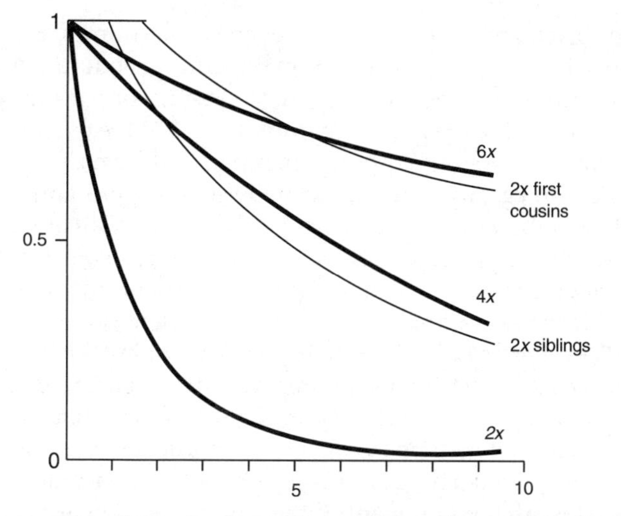

Fig. 15.1 Comparative effects of various patterns of inbreeding on loss of heterozygosity. Selfing is represented by broad lines for diploids (2x), autotetraploids (4x) and autohexaploids (6x). Matings between siblings and between first cousins are represented by narrow lines.

Polyploids have the same patterns of chiasma variation as diploids, outbreeding species showing variation within populations and inbreeders exhibiting differences between populations. They also show an interesting spectrum of variation in pairing behaviour. Autopolyploids usually produce high frequencies of multivalents, where more than two homologous chromosomes are paired, whereas allopolyploids rarely produce multivalents because they have evolved from interspecific hybrids so that their chromosomes are only partly homologous at best. The genomes of most natural polyploids represent a spectrum from the complete homology seen in autopolyploids to the non-homology seen in extreme allopolyploids; they are known as segmental allopolyploids. In polyploids, the interaction between meiotic and breeding behaviour extends to multivalent frequency. Outbreeders form multivalents whereas inbreeders do not. Some outbreeders do not maintain the multivalents beyond the pairing stage but that is due to lack of sufficiently well-placed chiasmata. In some strict allopolyploids, the lack of multivalents reflects a lack of homology but in many it is due to a diploidizing system of genes which suppresses multivalent formation. Diploidizing systems have been studied in most detail in crop plants such as cotton, oats and especially wheat. Diploidizing represents a restriction on recombination between different genomes so that a segmental allopolyploid can remain a permanent hybrid in the face of continuous self-pollination. The bread wheat *Triticum aestivum*, for example, is an allohexaploid consisting of two A-genomes, two B-genomes and two

D-genomes. Each genome consists of seven chromosomes. Each chromosome has one homologous partner from the same genome and four partial homologues from the other two genomes. A series of pairing suppressor genes, particularly the PH gene on chromosome 5 in the B-genome, ensure that only fully homologous chromosomes pair and form chiasmata. In this way, any heterozygosity between partially homologous chromosomes is preserved even though *T. aestivum* is regularly self-pollinated.

Outbreeding polyploids not only form multivalents but they also exhibit variation in multivalent frequency, particularly within populations. Again, the best studied examples are hexaploid grasses, particularly Timothy *Phleum pratense* and the hairgrass known in Britain as Somerset grass *Koeleria vallesiana*. Some of this variation reflects differences in chiasma frequency, since multivalents require more chiasmata to hold them together at first metaphase than bivalents do, but some is independent of chiasma frequency and reflects phenotypic variation in pairing behaviour. This latter component shows differences between populations as well as within them. Evidently, selection is acting on multivalent frequency, favouring different levels of inter-genomic recombination in different populations (Callow and Parker, 1985).

15.3.1 PARTHENOGENESIS – DEGENERATION OF THE GENETIC SYSTEM

As we have seen, genes control all aspects of the sexual cycle. When these genes vary in outbreeding populations they can lead to selective responses such as changes in patterns of recombination. If they do not vary, as within inbreeding populations, there can be no response to selection. Outbreeders can thus be seen as long-term generalists whereas inbreeders are short-term specialists. Inbreeding is a secondary strategy, derived from outbreeding, giving short-term advantage but no long-term security. It evolves over many generations of selection, during which advantageous combinations of non-allelic genes become fixed in the homozygous condition. Even the more specialized systems, such as interchange heterozygosity in *Oenothera*, show evidence of gradual accumulation of interchanges over time. Genetic systems change more slowly than the genetic variation which they control; selection acts directly on the target genes, not on the control systems. The exception is when the whole control system breaks down. Once the sexual cycle is broken, recombination either ceases altogether or is reduced to a minuscule subsexual level. This is the situation of parthenogenesis (virgin reproduction). If meiosis is involved, the system is said to be automictic, otherwise it is apomictic. Parthenogenetic animals often occur in exclusively female populations, although some molluscs are hermaphrodite and parthenogenetic. Parthenogenetic plants are usually hermaphrodite; in the absence of genetic variation, there is no mechanism whereby male organs can be lost under selection. Pollination is sometimes needed to stimulate egg development, although no fertilization takes place – a situation known as pseudogamy. An analogous situation pertains in the triploid salamanders of the *Ambystoma jeffersonianum* complex in North America, where sperm from diploid males is required to stimulate the development of parthenogenetic eggs.

Parthenogenesis represents the ultimate restriction on recombination. It usually becomes established in extreme habitats where sexual reproduction is either costly or difficult. The gekko *Heteronotia binoei*, for example, inhabits arid inland regions of Australia. Populations of triploid parthenogenetic females occupy less hospitable habitats than those of the bisexual diploids (Moritz, 1984). In plants, the frequency of parthenogenesis increases with latitude and often also with altitude. A very clear example is provided by the southern couch grass *Agropyron scabrum* (Hair, 1956). This hexaploid species is outbreeding in the North Island of New Zealand, facultatively parthenogenetic in the northern half of South Island and obligately parthenogenetic in the extreme south. Several genera of grasses include parthenogenetic species in sub-Arctic regions, notably *Deschampsia*, *Festuca*, *Hierochloë* and *Poa*. Parthenogenesis is more widespread in brambles (*Rubus*), hawkweeds (*Hieracium*) and dandelions (*Taraxacum*). Mendel fortunately began his studies with peas (*Pisum*) before taking an interest in hawkweeds; had he attempted things the other way round, he would probably have given up!

Parthenogenetic populations present special problems to taxonomists. Each population can be morphologically distinct and, being non-sexual, reproduces true to form and is genetically isolated from its neighbours. Each isolated clone may therefore be given a separate name as an agamospecies, especially by taxonomists who tend to split species up rather than lumping them. Several thousand agamospecies have been described in *Rubus*, *Hieracium* and *Taraxacum*. Because parthenogenesis reproduces true to form, its effects are often confused with those of inbreeding.

Where sexual reproduction has broken down, the relative amounts of homo- and heterozygosity depend on the mechanism of parthenogenesis. In apomictic systems, all heterozygosity is conserved. The presence of meiosis in automictic systems, which would otherwise lead to a halving in chromosome number, is off-set either by fusion of identical haploid nuclei or by a failure in nuclear division so that the chromosomes divide but the nucleus does not: a process known as nuclear restitution. The timing of the fusion or restitution determines the degree to which heterozygosity is conserved or eliminated (Figure 15.2). In organisms such as the Australian grasshopper *Warramaba virgo*, where restitution replaces the premeiotic mitosis, the chromosomes enter meiotic prophase in pairs of pre-aligned identical products of replication. Although chiasmata are formed, they do not produce recombinants because there are no differences to recombine.

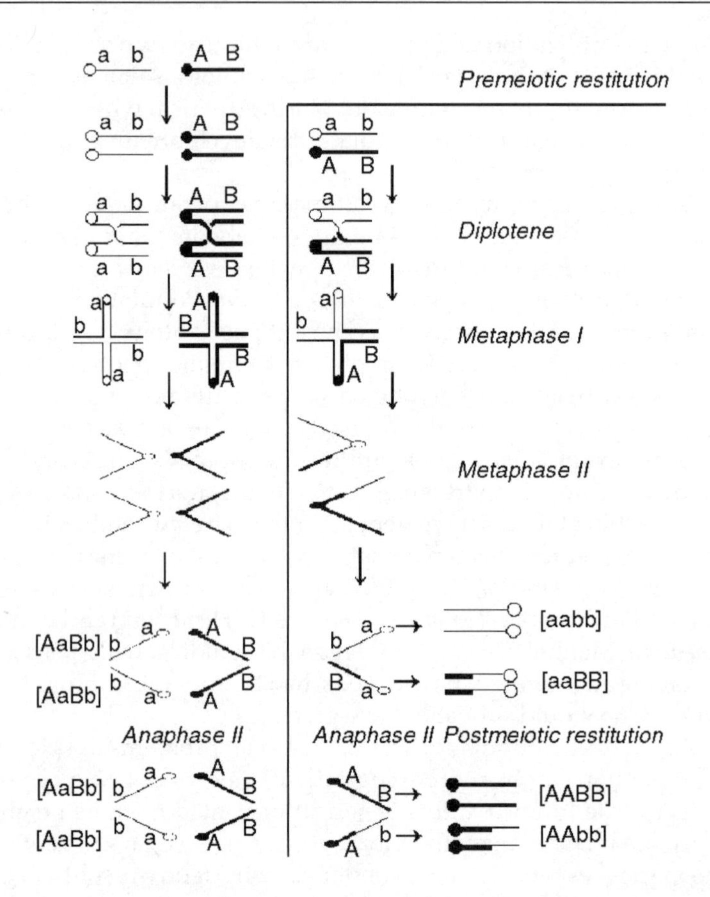

Fig. 15.2 Influences of the timing of mitotic restitution on the conservation or elimination of heterozygosity. Premeiotic restitution (left) leads to preferential pairing between the pre-aligned identical products of replication. Each gamete (genotypes in square brackets) therefore contains an intact copy of each parental chromosome so that all heterozygosity is conserved. Postmeiotic mitotic restitution (right) causes each gamete to contain two identical products of replication so that all heterozygosity is eliminated. Successive meiotic divisions are shown in the same orientation because meiosis is polarized in female animals and flowering plants and male meiosis has no consequence for parthenogenesis. Telocentric chromosomes are depicted for simplicity.

The identical chromosomes separate at the first division of meiosis and all heterozygosity between non-identical homologues is conserved. A similar system operates in triploid salamanders of the *Ambystoma jeffersonianum* complex. By contrast, postmeiotic fusion of identical haploid cleavage nuclei in scale insects (Hemiptera) and whiteflies (Homoptera) results in the complete and instant elimination of heterozygosity (Figure 15.2).

The timing of meiotic restitution also has critical effects on heterozygosity. In plants such as *Agropyron scabrum* and *Taraxacum officinale,* where restitution usually occurs at the first division, heterozygosity is conserved for segments close to the centromere but 50% of that in segments distal to a chiasma is lost (Figure 15.3). The limited amount of recombination possible in such systems is described as subsexual reproduction. If it is the second division which fails, as in the brine shrimp

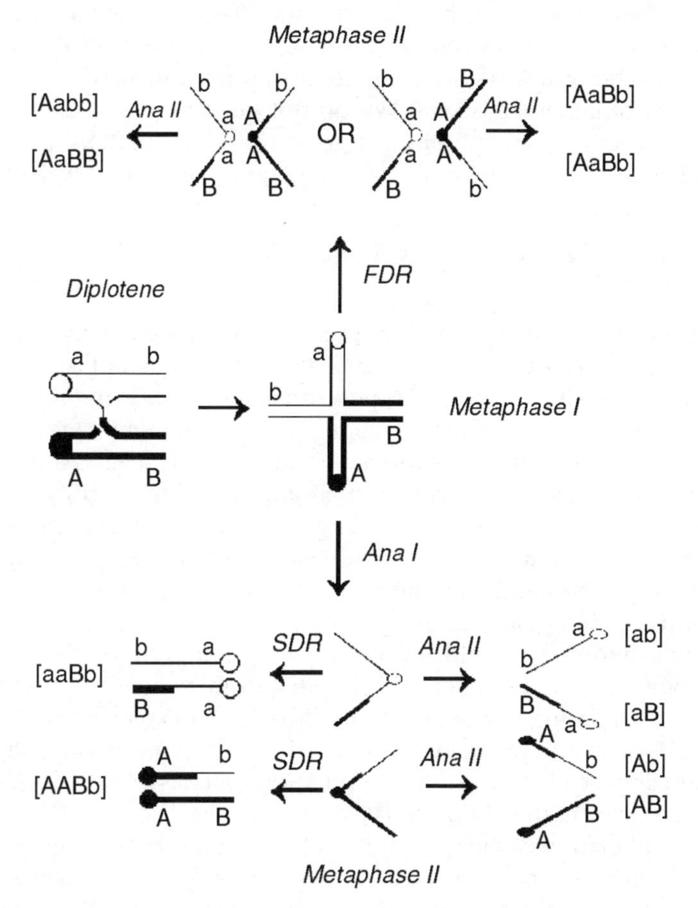

Fig. 15.3 Influences of the timing of meiotic restitution on the conservation or elimination of heterozygosity in chromosomal segments on either side of a single chiasma. The possible fates of a bivalent at metaphase I following normal meiosis (*Ana I* + *Ana II*), first division restitution (*FDR* + *Ana II*) and second division restitution (*Ana I* + *SDR*). Heterozygosity is conserved for genes (A/a) in proximal segments (between the chiasma and the centromere) by FDR but for genes (B/b) in distal segments (beyond the chiasma) by SDR. It is eliminated from proximal segments by SDR, although only one metaphase II cell may be affected, and reduced to one half in distal segments by FDR. The limited amount of recombination between genes in proximal and distal segments, following FDR, is known as subsexual reproduction. *Ana I*, anaphase I; *Ana II*, anaphase II; *FDR*, first division restitution; *SDR*, second division restitution. Other symbols as in Figure 15.2.

Artemia salina, then heterozygosity is eliminated from proximal segments but conserved in distal segments (Figure 15.3). Some plants, such as *Hierochloë odoratum* and *Taraxacum officinale*, show mixtures of first and second division failure at male meiosis (Ferris *et al.*, 1992). Where female meiosis has been observed, as in *Agropyron scabrum*, it seems to follow a similar course to male meiosis. Because selection acts on the target genotypes of the genetic system, rather than on the genes controlling variability, its action appears to be second hand. This distinction has led to arguments as to whether selection can act upon groups rather than individuals. Such arguments have largely arisen from misunderstanding of the nature of genetic systems. We do not need to imply that selection favours the group, or the ultimate good of the population, in order to appreciate the long-term nature of the evolution of genetic systems.

15.4 WHAT ARE THE MAIN CAUSES OF GENIC POLYMORPHISM?

The path from consideration of mutation rates takes us easily to complex levels of genomic organization. It appears that a number of issues such as the control of chiasma frequency, polysomic inheritance and so on can be understood in terms of genic selection. The question why the variation is there evidently requires an explanation at the genic level. At first sight, natural selection should favour a monomorphic, highly fit, genome. In reality, each individual harbours a large array of genetic variability. This has manifest advantages, because it ensures that populations can adapt very rapidly to new selection pressures as they arise. The suggestion that variability is present to ensure rapid adaptation is, however, untenable in Darwinian terms. It would imply that selection favours the long-term good, whereas natural selection operates from moment to moment to favour the offspring of the fittest individuals. An explanation of genetic variability is therefore essential to our view of the mechanism of evolution, and almost all other areas of population genetics relate to it. What can we conclude from the discussion?

One explanation of the variability is that it represents the genetic accidents which build up as a result of misreplication and intranuclear events. Some of it would be selected against, and indeed, there is evidence of a continuous process of elimination of deleterious mutant genes. This explanation depends, however, on the assumption that most genetic variants have no phenotypic consequence: they are irrelevant rather than beneficial or malign. The equilibrium level for irrelevant genetic variation depends on the balance between mutation and accidental loss, and can therefore be understood in terms of mutation and of loss, which is inversely related to population size.

It is very difficult to decide whether the observed variability can be reconciled with the evidence on mutation rate and population size. Influential geneticists such as Kimura and Nei have argued that it can, and that most of the variability at the molecular level is due to neutral mutations. Good agreement often depends on unverifiable assumptions, however, and the allele frequency distributions are usually compatible with other explanations involving selection. If the neutral explanation is not accepted then we have to explain variation in terms of selection of some sort. Some attempts start from trying to understand the behaviour of alleles at single loci. If selection on genotypes is invariant with frequency or density, then polymorphism implies heterozygote advantage or a nice balance between differing selection at different places or times. Field observation suggests that heterozygote advantage is rare. Balance between differing selection patterns at different parts of the habitat is, however, probably common. Frequency-dependent selection is, by observation, also prevalent, especially in relation to effects of predation and interactions between the sexes. Probably this explains a large fraction of the phenotypic variability.

It may not be justified to try to account for variability locus by locus. The fitness of an individual depends on an ensemble of genotypic effects, and some sort of interaction between them is inevitable. Many studies have been successful when the effects of loci were studied as if they were independent. In the sickle cell case, for example, all the discussion relates to the consequences of an alteration at a single nucleotide. On the other hand, in *Cepaea* explanations require that we consider colour and banding together, rather than one at a time. When we move to quantitative characters, it would be absurd to try to study the effects of segregation at a single locus in isolation.

One multi-locus explanation concerns the evolution of dominance and of heterozygote advantage. Since there is not much evidence of heterozygote advantage, concrete evidence that it has evolved is also lacking. On the other hand, super-genes of distinct loci controlling a suite of characters do suggest a fashioning of the genome by selection, especially in connection with visible polymorphisms such as those concerned with mimicry. The evidence from the structure of these super-genes almost certainly shows that they have been evolved by natural selection in the manner described.

It is tempting to try to understand multi-locus models using the same type of algebra as is applied to a single locus, but extended to two, three or more loci. Unfortunately the algebra quickly becomes extremely involved, and many assumptions have to be made. This approach is therefore of limited value. The alternative is to use the methods of quantitative genetics. Here, two types of model can be conceived to account for variability. One is the 'quantitative' version of the neutral allele model: genetic variance may be generated by near neutral mutations, balanced by random loss, and to some extent, by stabilizing selection. The other is the quantitative version of the balancing selection model. Loci interact to form the phenotype and selection for homeostasis favours an array of heterozygous loci. Analysis locus by locus may then be unable to detect the fitness effects which maintain the polymorphism. Some authors, such as J.F. Crow (1986) argue that the evidence favours the first explanation, rather than the second.

Bearing all these provisos in mind, is it possible to reach a conclusion as to how the polymorphisms are maintained? One of the difficulties in answering this question stems from the process of problem-solving itself. Scientists, as human beings, are competitive. They hope not only to perceive the truth, but to do so before, or more clearly than, their colleagues. One way of proceeding is to adopt a particular standpoint, a 'selectionist' or a 'neutralist' one in this case, and to advocate it for all possible situations, pushing the preferred model as far as the data will permit. Other investigators then do their best to contradict the conclusions drawn, and reinterpret the data according to their own preferred model. This is undoubtedly a very powerful method of getting at the truth – it is the one which is used in law. However, there is a danger of being persuaded by the skills of the advocate, rather than by the facts, and it may be that reality is just not as tidy as the models. Suppose that some systems are neutral, while others are mildly and yet others strongly selected. The arguments generated by adopting a particular standpoint may do more to confuse the issue than to clarify it.

Population genetics is not likely to gain any more from polarization of the issues in terms of neutralist and selectionist models. A spectrum of selection values exists, and varies through time and space. It is almost certain that many molecular variants are effectively neutral most of the time, and that divergence between taxa takes place as a result of accumulation of random substitutions. The most plausible general explanation of quantitative variation is that the alleles involved are usually functionally equivalent, and that the variability results from a balance of mutation and stabilizing selection.

Most polymorphic loci affecting the phenotype, on the other hand, are maintained by selection, although evidence for the underlying balance may be difficult to see. Suppose one allele is favoured at one end of an environmental cline, and an alternative allele at the other. The gene frequency might change from 99% at one end to 1% at the other, and the pattern be maintained by strong selection balanced by migration. Clines of this kind have been studied in mice of the genus *Peromyscus* in the USA, in which coat colour changes as one moves from areas of dark soil to light sandy ones and in the peppered moth in Britain where melanic frequency changes from urban to rural regions. If we examined colonies at the 50% point, no evidence of the dynamic nature of the balance is likely to be evident; only by disturbing the balance could we hope to see a compensatory change in gene frequency. In *Drosophila*, a build-up of alcohol in the larval medium is a potential hazard, and the presence of alcohol dehydrogenase polymorphisms is probably indicative of selection by this factor. Most of the time, alcohol levels may not be sufficiently high for the selection to operate, however, and the polymorphism depends on the right kind of selection operating for some of the time (this point is developed by Chambers (1988)). Again, the investigator will be hard put to find evidence of it. One feature which is noteworthy when models of linked arrays of populations are studied is that the expected duration of a polymorphism may be increased. The combination of selection and migration may prolong the survival of an allele to the point where a transient process due to directional selection, differing between different places, appears to be balanced.

If selective balance is difficult to detect in practice, the matching of genetic structure and phenotypic expression exhibited by super-genes shows that groups of genes have evolved to perform particular functions. Another phenomenon indicating the same kind of evolution is the canalization or homeostatic buffering, which occurs during the development of embryos and juveniles. These observations provide clear evidence for the importance of selection in fashioning the genome.

So-called neutral and selected alleles may therefore have more or less equal roles. It is probably fruitless to try to distinguish between them, and population genetics is the poorer for having left numbers out of the equations. If we go back to the original Darwinian view of evolution, we see a process which is continually modulated by density-dependent selection. The organization of the genetic material indicates that a wide range of different molecular configurations may lead to similar phenotypes. These phenotypes will be subject to selection when conditions are competitive or difficult, but could be effectively equivalent for long periods between. Since they are sometimes selected, it is not appropriate to treat them as neutral, but neither do they fit comfortably into one of the available categories of balancing selection. Explanations of polymorphisms should not be in the either/or categories in which we have tended to place them. A unitary theory may develop, which has as its first step the treatment of the fate of new mutants in terms of change in numbers, rather than in frequency. The problem with introducing numbers is that so many parameters creep in that it is difficult to generalize the results. The approach may be successful in the future, where it has had limited application in the past, if it can be developed in a communicable form in terms of computer simulations.

Another level is introduced when we look at the chromosomal mechanism for transferring genes from one generation to the next. Inversions or interchanges occur by accident to modify the process. They are probably usually disadvantageous and eliminated as they arise. As we have seen above, however, genic selection may favour a particular new arrangement, and rarely it may constitute a structural improvement. When interchanges or inversions do become incorporated, there is an opportunity for increased genic diversity. Mutation will accumulate to increase divergence between the alternative structural arrangements; inter-locus selection is facilitated, so that coadapted gene complexes arise. The behaviour of chromosomes plays an essential part in the explanation of genic diversity.

15.5 A NOTE ON STATISTICAL LIMITS

It makes a lot of difference to the theory if the average selection coefficient is 1% rather than 1 in 10 000, but it is very difficult to distinguish between these values experimentally. The odds seems to be against the investigator. To illustrate some of the problems, suppose that selection is acting on the three genotypes controlled by two alleles at a locus. Given the pattern outlined in section 7.5.3 we have

genotype	AA	AA′	A′A′
before selection	p^2	$2pq$	q^2
after selection	$\frac{w_1 p^2}{w}$	$\frac{2pq}{w}$	$\frac{w_3 q^2}{w}$

where $\overline{w} = w_1p^2 + 2pq + w_3q^2$. If we could sample the population before and after selection then p could be estimated from the data representing the population before selection. This value could then be used with the postselection data to estimate two selective values w_1 and w_3 relative to the third, w_2, which is here taken to be unity.

It is very rare, however, to have both sets of data. If only the post-selection frequencies are available, it is impossible to estimate p. The only course of action open would be to assume that the system is at equilibrium, so that p is the same before and after selection. It would then be possible to compare the observed frequencies with the Hardy–Weinberg expectation. A relative excess of heterozygotes over expectation would indicate heterozygote advantage if the assumption of equilibrium frequency is correct, otherwise it could indicate directional selection.

Suppose we know the frequencies before and after selection. Even then the problems are not over, because the next generation is formed after a round of gamete production and mating. The next generation can only be inferred from the quantities shown provided all genotypes have the same fertility and mating is at random. If either or both of these assumptions are incorrect then a further change will take place, and its effect may appear to indicate frequency-dependent selection of the zygotes.

In practice, it is often impossible to distinguish all three genotypes. In that case, an estimate could only be made from pre- and postselection data, or from a change in frequency over one generation, if an assumption is made about the relative fitnesses of the two genotypes which are indistinguishable. It often seems reasonable to assume that the fitnesses of these two genotypes are the same, but that is not necessarily the case, and may be the question at issue when a polymorphism is being investigated.

Bearing all these provisos in mind, what sort of sample size is required to detect a difference? To get some idea, we will consider the simplest situation. Suppose we catch samples of peppered moths in two successive generations and wish to test whether the frequency of the melanic form has changed. The melanic is dominant, but it is not necessary to know that to see whether a change has occurred. Suppose the first frequency is p_1 and the second is p_2, and the sample sizes are n_1 and n_2. The difference may be tested using the χ^2 (chi-squared) test. In this case

$$\chi^2 = \frac{n_1 n_2 (p_1 - p_2)^2}{n_2 p_1 (1 - p_1) + n_1 p_2 (1 - p_2)}$$

The frequencies are significantly different at the 5% level if χ^2 is equal to or exceeds 3.84. Now suppose for simplicity that we arrange to take the same sample size n in the two generations and instead of p_2 we write $p_1(1 - s)$, so that s is a selective coefficient for the melanic compared with the non-melanic form. We can now rearrange the χ^2 equation so as to calculate the value of n required to detect a selective coefficient s at the 5% level of probability. The rearranged equation is

$$n = \frac{3.84[2(1-p) - s(1-2p+sp)]}{s^2 p}$$

$$\text{or about } \frac{7.68(1-p)}{s^2 p} \text{ when } s \text{ is small}$$

The subscripts have been dropped because p refers to a single generation. Examining this equation we see that the sample size needed depends on the frequency, but more importantly, that n is inversely proportional to s^2. In the peppered moth, a selective advantage of 50% ($s = -0.5$) has been recorded. At that level of selection, the sample size required to detect a significant change when p is 0.5 is 35, an easy figure to achieve. If the selective advantage were 0.1, however, n would be 770, and for a 1% advantage, samples of 77 000 would be required. Apart from the labour involved, it would be physically impossible to catch so many peppered moths at one site in a season.

Various methods can be devised to minimize the difficulty, and results can be averaged over sites or over generations. There is a large literature on statistical testing procedures, much of it discussed by B.F.J. Manly (1985). Nevertheless, it is exceedingly difficult to find an experimental design which allows a small but important amount of selection to be distinguished from no selection at all.

15.6 SUMMARY

Prokaryotes and eukaryotes differ in the way in which they generate variation. Prokaryotes have: (1) the enormous populations necessary for rare mutants to occur and be selected on a regular basis, (2) haploid life cycles so that new mutants will be immediately expressed, (3) very short generation times and (4) reproductive systems with little or no recombination. Eukaryotes rely on sexual reproduction to generate variation through recombination. All aspects of meiotic and breeding behaviour are subject to integrated genetic control. These include the average level of chiasma formation and the localization of chiasmata. Diploid outbreeders represent the mainstay of eukaryotic evolution; they have the flexibility to adapt to changing environmental conditions. Restrictions on recombination may arise through inbreeding, chromosomal polymorphism, polyploidy or parthenogenesis. Each new system arises spontaneously and its future is determined by short-term selective advantages.

There has been a tendency to view genic variability either in terms of the neutral model or of selection. However, selective values probably range from negligible to very strong, and density- or frequency-dependent selection varies with ecological conditions. On average, there is probably more equivalence between alternative alleles and loci at the molecular level than at that of the phenotype. Selection at the molecular level is weaker and mutation is correspondingly more important as a contributor to polymorphism. Polymorphism resulting in genetic variance of biometrical characters probably arises from a balance of mutation and drift, only loosely constrained by stabilizing selection. At the other extreme, groups of loci which determine particular aspects of the phenotype have evolved under selection into super-genes, and there are many instances where the expression (dominance or recessiveness) of alleles has evolved. These phenomena show the pervasiveness and effectiveness of selection; they are important in increasing adaptation. In some cases, they result in polymorphism, although most selectively maintained polymorphisms arise from frequency-dependent selection or from a balance of selection and spatial heterogeneity.

15.7 FURTHER READING

Crow, J.F. (1986) *Basic Concepts in Population, Quantitative and Evolutionary Genetics.*
Gale, J.S. (1990) *Theoretical Population Genetics.*

Gillespie, J.H. (1991) *The Causes of Molecular Evolution.*
John, B. (1976) *Population Cytogenetics.*
King, M. (1993) *Species Evolution. The Role of Chromosome Change.*
Korol, A.B., Preygel, I.A. and Preygel, S.I. (1994) *Recombination Variability and Evolution.*
Mather, K. (1973) *Genetical Structure of Populations.*
Rees, H. and Jones, R.N. (1977) *Chromosome Genetics.*

PART THREE
Population genetics and evolution

Species formation and evolution

16

16.1 INTRODUCTION

In sexually reproducing higher organisms, the modern concept of a species is based on the capacity to interbreed. Within a species, there are likely to be subspecies (also called races), which are distinct forms living in different parts of the geographical range. Subspecies are interfertile, and represent adaptations to local conditions. A species composed of several subspecies is said to be polytypic. A fauna or flora usually contains many examples of polytypic species, each distinct from the next, and with little or no evidence that there has ever been genetic continuity. How then do new species evolve?

16.2 ALLOPATRIC SPECIES FORMATION

Three types of species formation may be distinguished. Probably the most common is allopatric speciation. Polytypic species often consist of subspecies separated by partial barriers to migration. For example, a small mammal may live in a series of valleys in a mountainous region, separated by mountain ranges and passes. If the barrier becomes more complete then there is an opportunity for divergence. The divergence may be accidental, due to stochastic fluctuation in gene frequency, or it may be the result of adaptive response to local conditions unhindered by migration of differently adapted individuals from other parts of the range. The adaptation may be to local physical conditions, or it may operate to reduce or avoid competition with other species. Either way, and even if random divergence has occurred, the result may be to build up differences between the subspecies which tend to form a barrier to interbreeding. Such mechanisms result in premating isolation if they tend to prevent union of the gametes. Examples would be divergence in flowering time between plants responding to different summer temperature conditions, or differences in courtship behaviour in birds. Postmating isolation results if the mechanisms lead to hybrid inviability after union has occurred. The classic example is the mule, the robust but infertile result of union between a horse and a donkey. If such processes continue for long enough, then the eventual result will be the formation of new species from the one which originally existed. There are many examples of similar but distinct pairs of species living in eastern and western parts of Europe which have probably evolved in this way during the Pleistocene when ice barriers were formed between Scandinavia and the Alps. Another fertile area to look for evidence of allopatric speciation is island archipelagos. Many examples exist, of which one of the most often quoted is that of the finches of the Galapagos Islands. The possible events which would lead to formation of new species as colonizing birds move infrequently from one island to another are shown in Figure 16.1.

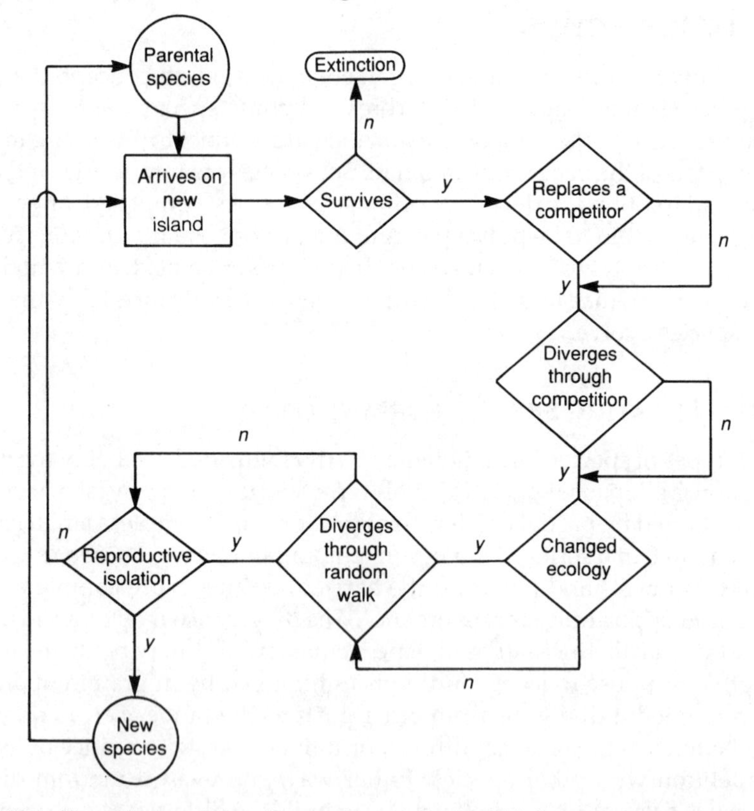

Fig. 16.1 Allopatric speciation. If animals such as land birds move from one island to another in a group of oceanic islands, there is an opportunity for divergence. An example is found in the Galapagos Islands, where 13 extant species and several fossil species in the subfamily Geospizinae have been generated from a single presumed ancestral species in 600 000 years. The range of possible events is outlined. For further details of the Galapagos finch case, see Grant (1986).

The characteristic of allopatric speciation is that geographical isolation precedes the other events which take place and permits divergence to develop. The isolation is an accidental occurrence. After separation, the new subunits may draw away from one another because of disruptive selection, but it is also possible that the changes which lead to reduced viability of subsequent inter-unit crosses are accidental. Only when they come together would selection consolidate the isolating process. The newly diverging stocks are likely to be selected towards different adapted modes but their opportunity to do so arises by chance. When two new species have developed from an ancestral species, there is also a chance that they may be able to coexist, increasing the species number in a region and probably narrowing the niches which they occupy.

16.3 SPECIES FORMATION FOLLOWING DISRUPTIVE SELECTION

The other major class of species-forming process is initiated by disruptive selection. There is disagreement about the relative importance of allopatric and non-allopatric processes overall, and their contribution to evolution in plants and animals. It is therefore useful to consider the disruptive selection process before describing the types of speciation involved.

16.3.1 DISRUPTIVE SELECTION AND ISOLATING MECHANISMS

Studies of the evolution of heavy metal tolerance in plants have proved very illuminating with regard to the action of disruptive selection in natural populations. Tolerance to heavy metals such as copper, lead and zinc is often measured as the length of root grown in a solution containing heavy metal as a proportion of the length grown in uncontaminated solution. Such measures are continuously varying characters controlled by polygenic systems. Heritabilities are high – of the order of 60% (Humphreys and Nicholls, 1984). The physiological basis of tolerance is likely to depend on a small number of genes, often specific to a particular metal, supported by a low level of general tolerance which may well be truly polygenic (Symeonidis *et al.*, 1985). Plants which are not able to tolerate heavy metals in contaminated soils are unable to survive and so the pollutant acts as a very straightforward selective agent. Studies of heavy-metal tolerance have been conducted on the monkey flower *Mimulus guttatus* and a number of grasses, particularly the common bent *Agrostis tenuis*, the sweet vernal grass *Anthoxanthum odoratum* and the sheep's-bit fescue *Festuca ovina*. One striking aspect of the variation in tolerance is the degree to which it follows the availability of the contaminant. In British populations of *F. ovina*, for example, the average degree of tolerance is positively correlated with the amount of extractable lead in the soil. Plants only seem to acquire as much tolerance as is needed. This is clear evidence for the role of natural selection.

Although mining for heavy metals can be traced back to neolithic times (*c.* 6000 years BP) in Europe, it did not reach its peak in Britain until the end of the eighteenth century. The evolution of tolerance in most British populations has therefore probably taken place within the last 200 years. When mining activity takes places within a locality dominated by non-tolerant species, the local populations immediately become exposed to disruptive selection. Plants exposed to the contaminant are under strong selective pressure for tolerance whereas those remaining in the uncontaminated zone are free from this selection regime. Moreover, plants in uncontaminated zones are under their own distinctive selection pressures, namely those of competition. This situation is made particularly clear by a classic study of a population of *Agrostis tenuis* (*A. capillaris*) interrupted by a copper mine at Drws y Coed in North Wales by McNeilly (1968). Drws y Coed is in a narrow glaciated valley, orientated in an east–west direction and exposed to strong westerly winds. McNeilly showed that the distribution of copper tolerance amongst plants sampled *in situ* was correlated with the degree of contamination in the soil. By contrast, non-tolerant plants downwind of the mine received pollen-carrying genes for tolerance so that their progeny had higher levels of tolerance than they did. Although not as extreme as that for tolerance around the mine, there was selection against tolerance in the uncontaminated zone. Tolerant individuals are uncompetitive. This is not a disadvantage on contaminated soil because the heavy metal will have killed off the opposition but it becomes a severe disadvantage when the contaminant is not present. Indeed, some uncompetitive species have evolved high degrees of tolerance and are more or less restricted to mining areas, not because they require the heavy metal for growth but because the contaminant keeps competitors at bay. Well-known examples in Britain include the Alpine penny-cress *Thlaspi alpestre* (*T. caerulescens*) and the vernal sandwort *Minuartia verna*.

In the population of *A. tenuis* at Drws y Coed, there are therefore two subpopulations exposed to different and incompatible selection regimes. Under such circumstances, any mechanism which prevents gene flow between the two subpopulations will be itself at a selective advantage. At Drws y Coed, the *A. tenuis* on the margins of the contaminated zone flowers about 1 week earlier than that in the centre of the contaminated zone and, critically, 1 week earlier than its neighbours in the uncontaminated zone. Similar observations have been made of *Anthoxanthum odoratum* around a mine boundary at Trelogan in North Wales. Both sets of observations provide excellent examples of the evolution of premating isolation in response to disruptive selection (McNeilly and Antonovics, 1968).

At face value, premating isolation is obviously more efficient than postmating isolation, since it does not involve the wastage of gametes, especially eggs which are costly to produce. It is therefore unlikely that a system of postmating isolation will evolve once premating isolation has become established, whereas postmating isolation could easily become superseded by a system of premating isolation. There are, however, some caveats which must be attached to this conclusion. First, because premating isolation is usually due to selection for different behavioural genotypes, it may well vanish if such selection lapses. A second problem is that premating isolation is not always effective. This is true of some animals, such as lizards in the genus *Anolis* where *A. aeneus* often competes for mates with *A. trinitatus*, but it is particularly true of plants where differences in mean flowering time do not preclude interspecific crossing which can even lead to the formation of extensive hybrid swarms, especially if the hybrids are fertile. Fertile F_1 hybrids can provide a bridge through which genes can flow between otherwise isolated species. Such introgressive hybridization is well known in perennial herbs such as the water avens and wood avens (*Geum rivale* and *G. urbanum*) and the white and red campion (*Silene alba* and *S. dioica*) and in trees such as the sessile and pedunculate oaks (*Quercus petraea* and *Q. robur*) and the downy and silver birch (*Betula pubescens* and *B. pendula*). In *Geum*, for example, although *G. rivale* flowers on average about 3 weeks earlier than *G. urbanum*, the flowering periods overlap by several weeks and both F_1 and backcross hybrids are common. In most localities, *G. rivale* is the most likely backcross parent so that gene flow is predominantly from *G. urbanum* to *G. rivale*.

Most systems of postmating isolation depend on chromosomal differences becoming established across the breeding barrier. Although these mechanisms are more costly than behavioural sources of premating isolation, they have the merit that they are rarely reversible. Moreover, their history can often be traced by meiotic analysis of inter-racial interspecific hybrids. Each breeding group may become homozygous for a different structural rearrangement so that any F_1 hybrids are heterozygous for several chromosome segments. Recombination in the F_1 hybrids gives rise to inviable gametes or unfit offspring. This is likely to be the situation where there is no obvious mechanical difficulty arising from structural hybridity. Examples include those paracentric inversions which are homozygous and monomorphic within species of *Drosophila* and polytypic Robertsonian fusions in races of the common shrew *Sorex araneus* and the black rat *Rattus rattus*. Populations of species separated by Robertsonian fusions sometimes exhibit relatively wide hybrid zones, within which the frequencies of the various fusions show clinal variation. Such wide hybrid zones testify to the relative ineffectiveness of Robertsonian fusion as a mechanism of genetic isolation. The best-known example of this situation is the gradient between the northern and southern races of the greater sand gerbil *Gerbillus pyramidum* in the Levant.

Where structural hybridity gives rise to obvious mechanical problems, distinct structural homozygotes appear to be fixed in different breeding groups. Those karyotypes which do not favour polymorphism, because they give rise to infertility in structural hybrids, are the very ones which are favoured under disruptive selection. Thus translocations other than Robertsonian fusions can cause postmating isolation, especially if they arise in acrocentric chromosomes so that the structural heterozygotes form multiples which are likely to be in adjacent rather than alternate orientation (Figure 16.2(b)). Whereas the parents show no evidence of structural abnormality or irregular pairing at meiosis, the F_1 hybrids contain multiple associations which are diagnostic of translocation heterozygosity. Relatively few examples have been described, possibly because their detection requires meiotic analysis of interspecific hybrids. The best-known examples are two subspecies of the grasshopper *Eprecocnemis plorans* and species of the plant genus *Crepis*, notably

PACHYTENE | METAPHASE I | ANAPHASE II

(a) Pericentric inversion | *Caledia captiva* | Moreton × Torresian

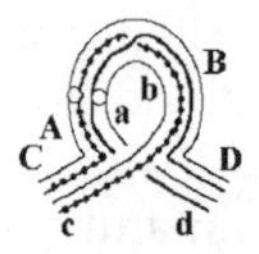

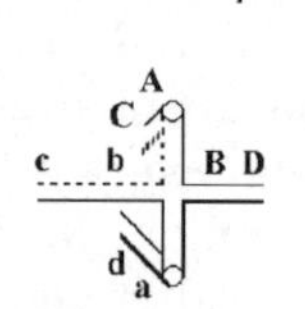

A
C B D
C b c
A
B
a D
d
d a b c

(b) Interchange *Crepis fuliginosa* × *C. neglecta*

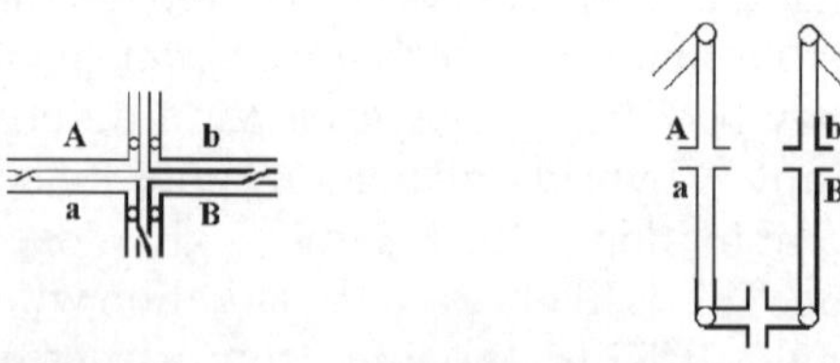

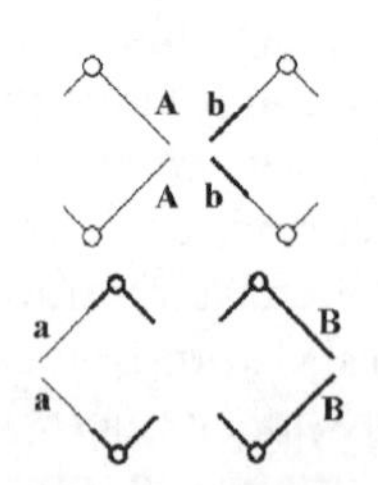

Fig. 16.2 Chromosomal mechanisms of postmating isolation. Here the opposite conditions apply to those which promote chromosomal polymorphism and fertility (cf. Figure 13.1). (a) Each large pericentric inversion causes a 50% drop in fertility in the heterozygotes whenever a chiasma forms within the inverted segment. In contrast to the equivalent situation in paracentric inversion heterozygotes, the recombinant chromatids which show duplication/deficiency for genes in distal segments (C/c, D/d) are not structurally abnormal and so are free to enter eggs as well as sperm. (b) Interchange heterozygotes show almost complete sterility when acrocentric chromosomes are involved, especially if they form multivalents, because the interchange multivalents are usually in adjacent orientation and so most gametes will show duplication/deficiency for non-homologous segments (bold and fine lines) and hence be inviable.

C. fulginosa ($2n = 6$) and *C. neglecta* ($2n = 8$). Detailed study of another genus in the same family (Asteraceae) has shown *Hypochoeris glabra* ($2n = 10$) and *H. radicata* ($2n = 8$) to differ by as many as three translocations, each apparently with break points very close to the centromere.

Pericentric inversions are often identified with postmating isolation, especially where the inverted segment is large enough to form a chiasma. Under such circumstances, each inversion for which the hybrid is heterozygous will cause a 50% drop in fertility (Figure 16.2(a)). The effectiveness of pericentric inversions as barriers to gene flow is demonstrated by the narrow hybrid zones which arise between species which come into secondary contact. An excellent example of this is the 1 km wide chromosomal hybrid zone 250 km in extent which includes populations of the Moreton (southern) and Torresian (northern) races of the Australian grasshopper *Caledia captiva*. This very narrow hybrid zone contrasts markedly with the distributions of Moreton mitochondrial and ribosomal DNA markers and at least four allozymes which extend as much as 400 km north of the hybrid zone (Arnold *et al.*, 1987).

16.3.2 PARAPATRIC SPECIES FORMATION

Examples like those discussed above, where distinct taxa are constant in form over a wide geographical area but are separated by a narrow zone of intermediate individuals are termed parapatric. Often it is difficult to decide whether the pattern represents a single species with distinct subspecies or a group of species which undergo partial hybridization. A striking example among European animals is the hooded crow/carrion crow complex. These two birds have characteristically different plumage patterns. Their ranges do not overlap, and they are separated by a narrow hybrid zone which follows no known environmental barrier and fluctuates in position from time to time. In house mice there are similar patterns in coat colour and molecular variants, including mitochondrial DNA. One explanation for these patterns is that allopatric speciation has occurred, and that the resulting taxa have rejoined. The alternative, parapatric, explanation would be that the patterns indicate incipient speciation.

As outlined in the previous section, clines in gene frequency tend to be set up in areas where there is a progressive change in the selective forces acting on alleles at a locus. Sometimes the environmental change is sharp, for example in metal concentration in soil. Soil colour may change abruptly from pale to dark and affect the camouflage of animals like mice. Other environmental factors, such as average temperature or humidity, change progressively. In either case, a relatively sharp transition in phenotype frequency may occur. If there is an area of low density somewhere along the cline then the gene frequency transition is likely to be located there, so that we have a pattern of patches of high or low frequency of alleles at particular loci connected by a quite narrow transition zone.

Within each patch, an individual who mates with another having the genotype favoured in that patch, produces fitter offspring than one mating with an individual from the other patch. There is therefore an opportunity for selection to favour genes for assortative mating. If this selection is effective, the steepness of the cline in the transition region will increase, and eventually complete reproductive isolation could be attained. Joseph Felsenstein has pointed out that the likelihood of such selection successfully resulting in speciation depends on the type of response required of the locus which reduces interbreeding. If it is the same on both sides of the transition, the chance is greater than if it is in opposite directions. For example, selection for reduced migration distance might assist steepening of the cline and would be equally effective at either extreme. On the other hand, change of flowering time would require early flowering at one extreme and late flowering at the other. In that case, linkage distance between the two loci, and formation of linkage disequilibrium, come into the equation and the appropriate conditions are less likely to be achieved. The parapatric speciation model predicts the fractionation of a polytypic species into a group of distinct species occupying different geographical areas. Ultimately, movement would be possible, so that some of these species come to coexist.

16.3.3 SYMPATRIC SPECIES FORMATION

It is also possible that disruptive selection towards alternative optima leads to speciation within a single area. Guy L. Bush studied flies in the genus *Rhagoletis*, which court and lay eggs on the fruits of various trees. There are several groups of species, such as those which feed on various species of walnut (*Juglans* spp.) in America, where speciation has probably been allopatric. Bush drew attention to a group of four morphologically very similar American species which specialize respectively on hawthorn, apple, plum and cherry. Those found on apple and cherry were first recorded in the mid-nineteenth and mid-twentieth centuries, respectively. The question is whether they could be the outcome of progressive specialization of some individuals from the original gene pool on those particular resources, which has proceeded to the point of speciation without geographical isolation. A model of sympatric speciation by disruptive selection has to start with polymorphism. In this case, we can conceive of a population of fruit flies carrying alleles which favour larval growth on either apple or plum but not both. As discussed in section 9.4, polymorphism could then be maintained provided (1) numbers on the two types of fruit were regulated independently, and (2) there was a balance between selection favouring each allele on its appropriate resource and gene exchange between the subpopulations on the two resources.

If such a polymorphism arises then genes favouring assortative mating, or any other mechanism which reduces exchange, will be selected. An effective response to this selection would result in speciation. In the fruit fly example, courtship and oviposition take place on the fruit, so that loci affecting efficiency of larval growth on, and adult preference for, a particular type of fruit could be the two components required. Indeed, it is reasonable to imagine these two characteristics being affected by the same gene. If at least two loci are involved, and disruptive selection is required at the isolating locus as well as the resource-allocating locus, then just as in the parapatric case, it becomes less likely that the necessary genic pattern will be achieved. There are many examples where we can conceive of sympatric speciation taking place according to this pattern. Whether the process is common depends on how likely it is that the necessary balance of selection and migration and linkage disequilibrium will be set up.

16.3.4 POLYPLOID COMPLEXES AND OTHER ISOLATING SYSTEMS

Another way whereby genetic isolation may be achieved is through purely numerical chromosomal changes, involving the duplication of whole genomes rather than individual chromosomes. Such changes produce geometrical series of chromosome numbers known as polyploid complexes. Where the series arises within an erstwhile diploid species, it is said to consist of autopolyploids. Where it involves interspecific hybridization as well as polyploidy, the polyploids are classed as allopolyploids. In both cases the bearers of the new arrangement constitute a new gene pool. It may seem that chromosome rearrangement is analogous to allopatric species formation in that an accidental event ensures isolation, after which divergence may or may not occur. For the most part, however, these changes are disadvantageous when they first arise, so that they are only retained when there is a pre-existing pattern of disruptive selection. When it leads to speciation, chromosome rearrangement is part of a selectively initiated process. Allopolyploidy presents extensive opportunities for innovation and has proved a major factor in plant evolution.

The chromosome number of eukaryotes is usually described as either gametic (n) or zygotic ($2n$), according to the stage of the sexual cycle represented by the tissue examined. When referring to the flowering plants, we may need to consider a further category represented by the endosperm which often has three times the gametic chromosome number ($3n$). In most cases, the status of the tissue under study is not in doubt so that this classification creates little difficulty.

Comparisons between species, or even between individuals of a single species, sometimes reveal that the zygotic chromosome numbers follow a geometrical series. Consider the zygotic chromosome numbers obtained from species of the onion genus *Allium* (Table 16.1). The three lowest numbers ($2n = 14$, 16 and 18) do not represent a geometrical series but diploid species derived from different basic monoploid numbers ($x = 7$, 8 and 9). The monoploid number (x) is the true haploid number. It is the minimum number of chromosomes carrying a complete genome. Unlike the gametic (n) and the zygotic ($2n$) numbers, it cannot be deduced from a single observation but has to be inferred from comparisons of chromosome numbers in different species. The majority of species of *Allium* have zygotic chromosome numbers which are multiples of eight. Thus the gametic number in the leek ($n = 2x = 16$) is equal to the zygotic number in the onion ($2n = 2x = 16$). The two species with odd numbers of sets ($3x$ and $5x$) both reproduce asexually.

Polyploidy arises through chromosome division accompanied by a failure of cell division: a process known as restitution (Chapter 15). A single restitution nucleus has double the normal number of chromosomes. Restitution may occur at diploid mitosis, giving rise to tetraploid sectors of tissue, or at meiosis resulting in unreduced diploid gametes. Autotetraploid individuals can develop through mitotic restitution whereas autotriploids arise from the fusion of reduced and unreduced gametes of the same species. In autopolyploids, the ploidy level usually corresponds with the maximum size of chromosome multivalent seen at metaphase I of meiosis. Thus diploids produce bivalents, tetraploids form up to quadrivalents, hexaploids up to sexivalents and so on. High levels of autopolyploidy buffer populations against the effects of inbreeding but are usually associated with reduced fertility.

Polyploidy has accompanied the evolutionary divergence of more than half the total of flowering plant and fern species. It is therefore a very important mechanism of genetic isolation in plants. Polyploidy appears to be much rarer in animals. It is usually associated with parthenogenesis, indicating that new polyploid races of animals probably have difficulty in reproducing sexually. For example, while diploid populations of the

Table 16.1 Chromosome numbers in species of the genus *Allium*, illustrating a series of basic numbers ($x = 7,8,9$) and a polyploid series amongst those with a basic number of $x = 8$ ($2n = 2x, 3x, 4x, 5x, 6x$)

Allium species		$2n$	Ploidy	
$x = 7$				
A. ursinum	Ramsons	14	Diploid	($2x$)
$x = 8$				
A. cepa	Onion	16	Diploid	($2x$)
A. sativum	Garlic	16	Diploid	($2x$)
A. schoenoprasum	Chives	16	Diploid	($2x$)
A. carinatum	Field garlic	24	Triploid	($3x$)
A. porrum	Leek	32	Tetraploid	($4x$)
A. vineale	Crow garlic	40	Pentaploid	($5x$)
A. roseum	Rose garlic	48	Hexaploid	($6x$)
$x = 9$				
A. triquetrum	Three-angled garlic	18	Diploid	($2x$)

Australian gekko *Heteronotia binoei* are bisexual, populations of triploids ($2n = 3x = 63$) are exclusively female and parthenogenetic. None the less, excellent examples of multivalent-forming autopolyploids are to be found in the Amphibia, notably the South American frog *Odontophrynus americanus* ($2n = 4x = 44$). This species is not only fully sexual but also, of course, sexually dimorphic.

Often the sterility of a diploid interspecific hybrid can be overcome by polyploidy because each chromosome then has an identical pairing partner rather than one which is at best only partially homologous. Such allopolyploids are distinguished by regular bivalent formation and can give rise to completely new true-breeding species which are genetically isolated from both parents. An example, which has arisen spontaneously in cultivation, is *Primula kewensis* ($2n = 4x = 36$), an allotetraploid of *P. floribunda* ($2n = 18$) and *P. verticillata* ($2n = 18$). Here, the genomic constitution of the new species may be written as FFVV, each capital letter representing the basic monoploid set ($x = 9$) of one of the species.

Similar events have produced new species when chromosomally and geographically isolated species have come into secondary contact, often as a consequence of human activity. A classic example is the new species of cord-grass *Spartina anglica* ($2n = 4x = 122$) which is an allotetraploid of the male-sterile interspecific hybrid *S.* × *townsendii* ($2n = 2x = 61$) between the native European diploid species *S. maritima* ($2n = 2x = 60$) and the North American diploid *S. alterniflora* ($2n = 2x = 62$). *S. alterniflora* is thought to have arrived with imported timber and the hybrid *S.* × *townsendii* was first recorded in Southampton water in 1878. *S. anglica* combines the vegetative vigour of the hybrid with full sexual fertility. Like, *P. kewensis*, this is a true-breeding, genetically isolated form which is a new species by any criterion. In Britain, it is now more widespread than the native species from which it arose, whereas the American parent has not spread beyond the Southampton area. A recent study of chloroplast DNA markers, which are maternally inherited, has shown that *S. maritima* is the female parent of *S. anglica* and *S. alterniflora*, the male.

Another well-known European example of a natural allotetraploid is the common hemp-nettle *Galeopsis tetrahit* ($2n = 4x = 32$) which is an allotetraploid of the F_1 hybrid between the diploid species *G. pubescens* ($2n = 2x = 16$) and *G. speciosa* ($2n = 2x = 16$). All three species are found on the mainland of Europe but only *G. tetrahit* and *G. speciosa* are found in Britain and Ireland. The large-flowered hemp-nettle *G. speciosa* is scattered but rather local in occurrence, especially in Ireland. By contrast, *G. tetrahit* is abundant and widespread. Evidently, as in *Spartina*, allopolyploidy has greatly extended the range of the original diploid species. Similarly, the geographical ranges of polyploid species of *Primula* and *Rosa* are much more extensive than those of the diploid species, particularly at high latitudes in the northern hemisphere. Extension during inter-glacial periods is bound to be greatly reversed once the ice returns, raising the interesting possibility that fresh cycles of polyploidy are initiated at each inter-glacial interval. Where the new polyploids exploit novel habitats at higher altitudes than those occupied by the ancestral diploids, as in *Buddleia* and *Rhododendron*, their geographical range is often much restricted.

Unequal crossing-over, duplication and gene conversion sometimes result in replication of identical DNA sequences within the individuals of one stock, which are different from those of another part of the same species. Such molecular divergence, known as concerted evolution, also gives rise to genic imbalance in hybrid individuals and it is possible that it plays a significant part in speciation.

16.3.5 CENTRES OF DIVERSITY

Many of our most valuable crop plants are polyploid including cotton (*Gossypium hirsutum*, $2n = 4x = 52$), oats (*Avena sativa*, $2n = 6x = 42$), potatoes (*Solanum tuberosum*, $2n = 4x = 48$), sugar cane (*Saccharum officinarum*, $2n = 10x = 80$) and wheat (*Triticum aestivum*, $2n = 6x = 42$). The Russian geneticist N.I. Vavilov made a detailed study of the origins and geographical distributions of cultivated plants. As a result of his research, he came to the conclusion that the world's cultivated plants arose chiefly around a small number of regions which Vavilov called centres of diversity. Not surprisingly, they were associated with sites of early civilization. Centres of diversity have now been recognized in many regions of the world. The nearest to Europe is known as the Fertile Crescent which runs from Palestine northwards through Syria and eastern Turkey, westwards to northern Iraq and then southeastwards along the rivers Tigris and Euphrates to the Persian Gulf, through the ancient region of Mesopotamia which is now Iraq. Many important crop plants evolved in the Fertile Crescent, along with early systems of agriculture which allowed the erstwhile hunter-gatherers to settle and build permanent settlements. Examples include cereals such as barley and wheat and pulses such as peas, beans and lentils.

The most important crop plant to have evolved via polyploidy is the bread wheat *Triticum aestivum*. Evolution occurred under agricultural conditions but the events were not consciously controlled by man. Wheat is the most valuable crop to have emerged from the Fertile Crescent and is likely to have been the deciding factor governing the earliest human settlements. Bread wheat is a hexaploid species with the genomic composition *AABBDD*. Its history has been traced via a combination of archaeological remains, polyploidy and breeding experiments. This is a tale of inadvertent selection by the earliest farmers (*c*. 7000 years BP). The earliest wheats to be cultivated were the diploid Einkorns (*T. monococcum*), with genomic composition *AA*, and the allotetraploid Emmers (*T. dicoccum*), with composition *AABB*. The genomes are likely to have originated in wild goat grasses (relatives of wheat) in the genus *Aegilops*: *A* from *Ae. aegilopoides* and *B* from *Ae. speltoides*. The Emmer wheats were larger and more prized than the Einkorn wheats so that their range extended westwards into what is now Syria. Here, hybridization this time with *Ae. squarrosa* (*DD*) followed by doubling resulted in the hexaploid *T. aestivum* which is now one of the three most important crop plants in the world (along with rice and corn).

16.3.6 THE GARDEN ROSE

Some cultivated plants have a chequered history, involving polyploidy, geography and trade. They reflect relatively recent natural selection following phases of dispersal and geographical isolation. Garden roses provide a good example. Although the genus *Rosa* spans a wide range of polyploidy from diploid ($2n = 2x = 4x$) to octaploid ($2n = 8x = 56$), only the diploids and tetraploids and their triploid hybrids have featured significantly in cultivation. The diploids have come from the Far East and the tetraploids from Europe and central Asia. The eastern diploids were delicate, tall or climbing plants with attractive and pleasantly scented flowers. By contrast, the tetraploids were robust bushes with many-petalled flowers. Although crossing a diploid with a tetraploid normally results in a sterile triploid, occasional unreduced gametes (especially eggs) from the diploid parent must have resulted in some fertile tetraploid hybrids.

It is very likely that roses were selectively bred for centuries within Europe, India and China. The first contact between the eastern and western groups seems to have come via the ancient silk routes between the civilizations of the Middle East and those of India and China. Damask roses have been known since Minoan times (*c.* 3000 years BP) and they take their name from the city of Damascus. They represent crosses between the musk rose (*Rosa moschata*) and the bush rose (*R. gallica*). Damask roses were highly prized because of their exquisite scent and slowly spread westwards, finally reaching England during the reign of Henry VIII. From Tudor times until the early nineteenth century, the Europeans (notably the Dutch) concentrated on breeding bigger and bigger flowers and producing moss roses and such monstrosities as the 'hundred-petalled rose'.

The noisette rose was produced in Charleston (USA) in 1802 directly from eastern diploid stocks, by crossing *R. moschata* with members of the gigantea group from China, known as *R. chinensis*. Further crosses between noisette roses and *R. gigantea* resulted in the delicate tea roses of the 1830s. These had exquisite flowers but were not very hardy. Meantime, another series of crosses was underway in Europe. These began with the Bourbon roses produced on the Île de Bourbon, when damask roses were crossed with *R. chinensis* in 1817. Further crosses of this type, in

France and in Britain, produced a series of hybrids known collectively as the hybrid Chinas and later the hybrid perpetuals. These roses were more attractive than the old moss roses of the eighteenth century but more hardy than the eastern diploids. From 1867 onwards, tea roses were crossed with hybrid perpetuals to produce hybrid tea roses which form the basis of most modern cultivars. Other groups have since been incorporated, bringing their own distinctive qualities: *R. multiflora* (many-flowered), *R. wichuraiana* (climbing) and *R. foetida* (yellow-flowered). Evidently, roses have been highly prized for several millenia, purely for their aesthetic properties. A simplified summary of this complex history is given in Table 16.2.

16.4 SPECIES FORMATION – CONCLUSION

In this very brief reference to a vast subject, we have identified three patterns of speciation – allopatric, parapatric and sympatric. They differ in the geographical scale on which speciation takes place. Allopatric speciation requires geographical isolation, after which divergence takes place. Parapatric speciation assumes partial separation of subspecies adapted to different conditions, which then move to complete separation. Sympatric speciation would separate individuals into different niches within a single habitat. We have also identified two types of dynamic process responsible for species formation. The allopatric model depends on accidental separation, after which natural selection will modify the resulting taxa independently. Parapatric and sympatric speciation depend on disruptive selection accompanied by selection for reproductive isolation. In these two models, natural selection is central to the species-forming process.

Chromosomal reorganization and polyploidy are both mechanisms whereby reproductive isolation may be effected or enhanced. Interspecific F_1 hybrids are usually heterozygous for at least one structural change, often involving either a pericentric inversion or interchange. Most postmating isolation has a chromosomal basis; polyploidy is the major source in plants but its role in animal evolution is less obvious. Systems of premating isolation, based on behavioural differences, are less costly than those of postmating isolation but they can be less effective and more easily reversed. The narrowness of hybrid zones often testifies to the effectiveness of the chromosomal barrier, even when nuclear and cytoplasmic genes have introgressed well into foreign territory.

Speciation is sometimes, but not always, directly involved with the process of increasing adaptation. Many workers would argue that allopatric species formation is the most common type. If so, species are continually coming into being by chance, to provide sets of isolated gene pools on which the evolutionary process may work.

Table 16.2 Simplified diagram illustrating the history of the garden rose up until 1950

Eastern: $2x$		$3x \rightarrow 4x$	Western: $4x$	
Tall	Intermediate or variable	Bush	Bush	
R. Moschata Musk rose Himalaya	*R. moschata* x *R. chinensis* ↓ Noisette rose (1802–1880) (Charleston, 1802)	*R. chinensis* x *R. damascena* ↓ Bourbon roses (1817–1870) (île de Bourbon, 1817)	*R. gallica* x *R. moschata* ↓ *R. damascena* Damask rose (*c.* 3000 BP)	*R. gallica* Bush rose S. Europe
		R. chinensis x *R. gallica* ↓ Hybrid China roses (1820–1860)	*R.damascena* → *R. centifolia* Cabbage R., 1580–1710	
R. gigantea Tea rose China, Burma	*R. gigantea* → *R. chinensis*		*R. damascena* → *R. muscosa* Moss rose	
	R. gigantea x Noisette R. ↓ Tea rose (1830–1900)	*R. gallica* x H. Chinas *R. gallica* x *Bourbons* ↓ Hybrid perpetuals		
R. multiflora Many-flowered rose China, Japan	*R. chinensis* x *R. multiflora* ↓ *R. polyantha* (Dwarf: 1873–1936) (Climbing: 1881–1910)	Tea roses x H. perpetuals ↓ Hybrid tea roses (1867–1950)		
		R. polyantha x H. tea ↓ H. polyantha (1920–1950)	*R. foetida* x Hybrid tea ↓ *R. pernetiana* (1898–1920)	*R. foetida* Austrian briar Persia?
R. wichuriana Climbing rose China, Japan	*R. wichuriana* x Tea rose ↓ Hybrid wichuriana (1890–1950)	H. wichuriana x H. tea ↓ H. wichuriana (1890–1950)		

After Darlington (1973).

16.5 TWO GENETIC MODELS FOR SPECIATION

In the above descriptions, it is suggested that fitness depends on combinations of alleles at numerous different loci. The genetic homeostasis argument implies the same thing. We should expect to find that chromosomes accumulate particular sets of alleles which work well together, while other chromosomes have alternative sets. Individuals with different chromosome sets, or haplotypes, may then exhibit multiple heterozygote advantage. Loci on other chromosomes may then be fixed for particular alleles because they work well with the polymorphic complex concerned. Such a system, where there is a high degree of epistatic interaction between loci, is said to be coadapted. A lot of discussion of the possible course of evolution concerns the relative importance of adaptation. It is an essential feature of some hypotheses about the conditions which would promote or retard evolution.

16.5.1 SHIFTING BALANCE

Sewall Wright developed the theory of adaptive topography introduced in section 9.2. If two loci interact to determine the mean fitness of a population, the topography is represented by a surface in three dimensions, representing $\overline{w}$ and the allele frequencies at the two loci. For n loci, the $\overline{w}$ topography is a surface in $n + 1$ dimensional space. As n increases, it becomes theoretically possible to have an enormous number of different adaptive peaks, representing different levels of adaptation, separated by adaptive valleys and low points. In other words, many different combinations of allele frequencies at the loci concerned may lead to local adaptive modes, and some of these will represent higher mean fitness values than others. Ideally, the population should move to the peak representing the highest possible mean fitness, but the contours of the topography make it probable that many will be trapped at local, lower maxima. Sewall Wright then posed the question, given that genomes are organized in the way described, what circumstances would be most likely to promote evolution? We would not expect to see adaptive responses in small isolated populations, because although they may be subject to strong selection there would be little genetic variability on which it could act. Very large populations would possess high levels of genetic variability but would suffer the stultifying effect of a large N_e on the ability to explore the adaptive topography. The ideal would be a situation where there was genetic diversity and small population size, so that gene frequencies could be subject both to selection and to random variation. These favourable circumstances could be achieved if a species were polytypic and divided into numerous small semi-isolated subspecies, each adapted to slightly different conditions. Where several come together, small populations would have the benefit of genes from a number of gene pools and the chance to move to a new, higher adaptive peak. If a much better genetic organization was achieved than in the surrounding populations, then there would be an opportunity for migration to spread the new type through the species. Evolutionary novelties would therefore tend to crop up in geographically complex but central parts of the range.

16.5.2 PERIPHERAL DIVERSITY

The shifting balance theory is an interesting derivation from our picture of elementary gene frequency change. We sometimes see geographical distributions of apparently advanced and primitive forms of species which fit the prediction, but that does not necessarily mean that they arose in this way. Ernst Mayr produced an alternative explanation which suggests that the peripheral populations are really the evolutionary novelties. The argument runs as follows.

Many examples are known where a species is relatively invariant over large areas of diverse environments, but has distinct forms in isolated populations around its margins. One of Mayr's examples was the distribution of kingfisher species and subspecies on New Guinea and nearby islands. According to this argument, the species does not vary on the mainland because of the homogenizing effect of migration, while isolation has allowed the marginal island forms to evolve. Now, simple selection, migration theory suggests that selective values would easily reach levels where they would override the effect of migration if local adaptation to differing environments was selected. Why, then, does migration have such a marked effect?

The answer lies in the existence of coadaptation. A balance of alleles at different loci tends to build up, and migration upsets the epistatic combinations which have evolved. If there is a continuous low level of migration, the result is likely to be a widespread coadapted complex, the 'mainland' species. By the same token, if isolated island populations are started by a few individuals they will carry only a small subset of the gene pool. If increase in numbers follows successful establishment then new mutations will be incorporated into the population, selection will be somewhat different from that acting on the mainland population and there is a chance that a new coadapted gene complex will be built up. Mayr described this bottle-neck, followed by divergence and establishment of a new integrated gene pool, as the Founder Effect. There is then a chance that the island type will reinvade the mainland and replace the older population.

We therefore have two theories about the most likely conditions to promote evolution. Both are based on simple population genetic ideas, but they differ in their conclusions. According to Mayr, coadaptation develops very tightly organized combinations of genes, isolation is essential before genetic change can take place, and when it does so, radical new combinations are born. Genetic drift as a continuous process is irrelevant, but the non-recurrent founder events are all-important. The adaptive topography model is also a description of coadaptation, but for Wright the degree of interdependence of loci is less strong. Isolation must only be partial, and gene dispersion in small semi-isolated populations is an essential part of the process.

Both models were put up to be tested against the evidence. So far as biogeographical data go, examples can be found to support each theory. So far as coadaptation is concerned, there is no evidence for the degree of integration implied by the Mayr model. There are cases where linked groups of loci have evolved an epistatic disequilibrium, but these are often multi-gene families or super-genes such as those controlling mimetic polymorphisms in butterflies. More extensive gene combinations are found in the inversion polymorphisms in *Drosophila* species. These are true coadapted sequences, and the net heterozygote advantage which they exhibit has sometimes been demonstrated to break down when genes from different populations are mixed. Nevertheless, integration at this level seems to be the exception rather than the rule. If coadaptation was as important as the models suggest, there should be many experimental results which show a marked loss of fitness as a result of mixing of geographically isolated or differently selected populations. Such evidence does not exist. Simple population genetics works because the effects of loci can be studied one by one. It must therefore be doubted whether coadaptation has the importance assigned to it, certainly by the Mayr model and probably even by the Wright model.

16.6 STATIC AND CHANGING PATTERNS

Whether or not either theory describes the evolutionary process, they have exerted a great deal of influence on the way people think about it. They also bring us back to another issue discussed above, namely the relation between species formation and evolution. The Wright theory appears to be a description of rapid phyletic evolution (sometimes called anagenesis). It would lead to long persistence of a lineage, perhaps with chronological species in the fossil record. The Mayr theory places emphasis on geographical speciation (or cladogenesis). While one focuses on the trunk of the phyletic tree, the other is concerned with its branches. In principle, another way of weighing up the evidence would be to examine the fossil evidence, to assess the importance of branching events in evolution.

Niles Eldredge and Stephen J. Gould concluded from a study of the fossil record that evolution frequently proceeds for long periods during which little happens (periods of stasis) followed by periods of rapid speciation and evolution. They argued that this pattern cannot be understood simply by extrapolating from elementary population genetics, but that some other processes must be involved, which can be thought of as a higher level of evolutionary process. One candidate would be chromosomal reorganization (polyploidy, aneuploidy, etc.) which would cause reproductive isolation so that selection and adaptation may take place within new, distinct gene pools. Another candidate would be coadaptation. If coadaptation was strong then constancy through time, as well as geographical area, would be expected, and rapid speciation would occur when circumstances were such that a shift from an old to a new coadapted mode could take place.

Support for this interpretation has been provided by P.G. Williamson. He studied fossil freshwater molluscs in African lake deposits. The sequences examined extended over four million years, and contained 19 species lineages in 18 genera, some gastropods and some bivalves, some sexually reproducing and some asexual. The common feature was a tendency for stasis followed by rapid changes in shell shape. In this context, 'rapid' means periods of 500–50 000 years.

It may be argued that this short period is a long time on the micro-evolutionary time scale, and that there is no need to invoke any explanations outside of elementary population genetics. In favour of the coadaptation hypothesis, however, it was observed that during periods of rapid change developmental homoeostasis, measured as variance of shell shape, declined. This would be expected if there was a major reorganization of a coadapted gene pool, although it could also indicate disruptive selection.

Whatever the interpretation of this particular example, the evidence against coadaptation is simply that genetic analysis has not revealed closely integrated complexes of the type required. We therefore have the same reservations as before. Constancy in time, like geographical similarity, is more likely to be due to stabilizing selection than to some sort of internal genetic inertia, and the jump to a new mode must indicate a new pattern of stabilizing selection.

16.7 SPECIES SELECTION

The question underlying the concept of punctuated evolution is whether there are further features of evolution of a higher level than have been discussed in the rest of this book. A formulation of the question current in the 1920s separated genetic changes into micro-mutations and macro-mutations. Changes such as that observed in the peppered moth and polymorphisms such as those involving sickle cell and C alleles were classed as micro-mutations. They occurred, fluctuated in frequency at random or were selected, as part of the variability within species, and polytypic species were evidence of diversification within species. Change from one species to another, however, required a macro-mutation. This point of view developed, in a proper scientific manner, to explain observations about the natural world. Variation was seen within species, but well-defined gaps appeared to exist between species. Major reorganizations in the form of polyploidy were known, and these can undoubtedly result in partial or complete reproductive isolation. Perhaps speciation normally occurred by quantum jumps, while population genetics was concerned only with fine tuning of the breeding group. Later, the neo-Darwinian synthesis, in which Mayr played a leading part, was an attempt to counteract this view.

Micro-evolution was seen as fashioning major new innovations. Species were considered to intergrade into each other if one examined the right part of the geographical range: the discontinuity between species is most evident when one looks at sympatric species which are selected to minimize niche overlap. However, if coadaptation is very well developed, and results in an inertia in response to selection until the system flips to a new adaptive mode, then the two-level model of evolution is to some extent revived.

As stated above, the evidence for this model is weak. Speciation itself, however, turns evolution into a two-level process. Whereas many loci are selected more or less independently of other loci, competition between species is selection of one set of integrated genotypes compared with another. A massive natural experiment occurred when North America and South America were joined through the isthmus of Panama during the Miocene, mixing the faunas of the two continents. For the most part, representatives of South American groups which moved north became extinct, whereas North American groups moving south survived, evolved and possibly contributed to the extinction of many South American taxa which subsequently disappeared. The classical interpretation of this asymmetry is that the northern taxa were competitively superior to those from South America. Among living mammals, we can get a glimpse of a similar phenomenon by examining the Australian marsupial fauna, from which the placental mammals have been almost excluded. In terms of genetics, the marsupials represent a set of integrated genomes programmed to reach a given set of adaptive peaks, while the placentals are differently organized, and can achieve different peaks, which often appear to have superior competitive power when they come into contact. Species selection, both between closely related species and at this broader level of competition between whole faunas, has probably played an important part in the evolution of life.

16.8 SUMMARY

New species may be formed allopatrically, in which case geographical isolation initiates speciation and precedes divergence and reproductive isolation. At the other extreme, sympatric speciation may occur. The initial trigger for this is disruptive selection, followed by the development of isolation which may be brought about by chromosomal reorganization. In parapatric speciation, divergent selection in different places sets up a cline in gene frequency which eventually steepens and results in reproductive isolation. In two of these processes, isolation precedes divergence while in the other two it succeeds it. If isolation comes first speciation is essentially an accidental process. The evidence suggests that the most common form of speciation is allopatric, and therefore accidental. Like mutation at the population level, isolation creates the conditions which may then lead to extinction of one or other of the types or to continued existence separately or together.

The shifting balance theory of Wright is based on the assumption that loci interact in such a way that numerous possible mean fitness maxima exist for different allelic combinations. The greatest increase in adaptation occurs if the frequencies move to the highest peak. Since there are points of low mean fitness between the peaks, this outcome is most likely when a population is small enough for some random fluctuation in gene frequency to take place. The theory rests on the idea of locus interaction, or coadaptation, and leads to a prediction that phyletic evolution is most likely to occur in small populations of mixed origin.

The peripheral isolation theory of Mayr assumes that coadaptation is strong enough to inhibit divergence between individuals in different geographical regions subject to the usual levels of migration. Complete separation of small colonies is therefore significant because it allows new coadapted combinations to form uninhibited by migration. Speciation is most likely in such isolates.

Both theories draw attention to the fact that locus interaction may be important in evolution. It may result in inertia in response to selection. Once species have been formed, interspecific competition selects one genome in favour of another, so that a higher level of evolutionary process may operate. All these ideas must be counted as speculative. They represent logical deductions from the theory and from the simpler types of evidence on which they are based, but they are difficult to confirm or deny on the basis of the fossil and geographical evidence of evolution.

16.9 FURTHER READING

Bradshaw, A.D. and McNeilly, T. (1981) *Evolution and Pollution.*
Claridge, M.F., Dowah, H.A. and Wilson, M.R. (ed.) (1997) *Species. The Units of Diversity.*
King, M. (1993) *Species Evolution. The Role of Chromosome Change.*
Maynard Smith, J. (1989) *Evolutionary Genetics.*
Mayr, E. (1970) *Populations, Species and Evolution.*
Mayr. E. (1976) *Evolution and the Diversity of Life.*
Skelton, P. (ed)(1996) *Evolution. A Biological and Palaeontological Approach.*
White, M.J.D. (1978) *Modes of Speciation.*
Williamson, M. (1981) *Island Populations.*
Wright, S. (1978) *Evolution and the Genetics of Populations, Vol. 4, Variability Within and Among Natural Populations.*

Species associations 17

17.1 GEOGRAPHICAL DIVERSITY

Evolving taxa each have a unique history. In the search for causes of evolution, it would be surprising if we found we could account for all the patterns which are observed. There may be a rational explanation for all features of a species or a species association, but each is likely to have characteristics that are consequences of past events which cannot be unravelled. At this point it is therefore interesting to ask whether the foregoing discussion results in a system which is sufficiently complete to account for the number and type of species in a given community. The archipelago of Madeira is an appropriate starting point.

17.1.1 GEOGRAPHY AND GEOLOGY OF MADEIRA

The group is volcanic in origin and lies 600 km from north Africa, the nearest land mass. Madeira is about 50 km long and 20 km wide. It rises to over 1800 m and has a cool damp climate at higher altitudes, while the southern coastal strip and eastern peninsula are sunny and dry in summer. The temperature varies little through the year. About 20 km to the south and connected by a submarine ridge are the three Deserta islands, dry, eroded and rising to 400 m. Porto Santo, the second largest island of the group, is about 40 km to the northeast. It is 10 km long and rises to 400 m. It is now dry and bare but was probably clothed in rich vegetation before colonization by man, which would have maintained a damper climate. There are several small offshore islets, the whole cluster being the relict of a once larger island. The ocean floor drops to more than 2500 m between Porto Santo and Madeira.

Madeira was created by a series of volcanic eruptions between three-quarters of a million and over 12 million years ago, some phases of this sequence probably almost totally destroying the pre-existing island. Porto Santo did not experience the later eruptions and is a more eroded island.

Running northeast from the group towards Portugal is a series of banks and sea mounts which may once have been islands. If so, Madeira is the youngest of a chain along which the flora and fauna were linked to Europe. Much, but not all of it, is derived from Europe. The islands are too far south to have been subject to the Pleistocene glaciations but would have experienced climatic fluctuations and sea-level changes synchronized with them.

17.1.2 ANIMALS WITH LOW ENDEMISM LEVELS

Among the organisms to be found now, there is a range of levels of diversity and endemism. There are no native land mammals, with the exception of bats, all others having been introduced by man. There appears to have been a mouse species present before the arrival of man; it is now replaced by the house mouse, *Mus musculus*. A single species of lizard, *Lacerta dugesii*, is abundant. It is endemic and shows subspecific variation in colour and scale patterns between islands. Like the endemic lizards of the Canaries, it is more vegetarian than mainland lizards of the same family. There are about 20 breeding species of land birds. Of these, the long-toed pigeon, *Columba trocaz*, Berthelot's pipit, *Anthus berthelotii*, and the plain swift, *Apus unicolor*, are endemic. Ten species have endemic subspecies. Among the rest, at least two, the greenfinch and the Spanish sparrow, are European species which have arrived and started to breed within the past 100 years. Residents tend to have rounder wings, less distinct sexual dimorphism and broader niches than those from Europe. Some species which would be expected, such as tits, are not present. There are 12 species of butterflies, of which one is an endemic species and four are endemic subspecies. One, the Indian red admiral *Vanessa indica*, has a disjunct distribution, occurring to the east no nearer than western Asia. Again, recent changes in distribution have occurred. The small white, *Pieris rapae*, arrived in the 1950s and may be replacing the large white, *P. brassicae*. The speckled wood, *Pararge aegeria*, may be replacing its endemic congenor *P. xiphium*. The milkweed butterfly, *Danaus plexippus*, arrives from time to time from America and breeds for a few generations before dying out. There are 1100 species of plants, of which 11% are endemic. Some of these are members of the montane subtropical rain forest, while others are adapted to a xerophytic existence on the rocky coasts.

17.1.3 ANIMALS WITH HIGH ENDEMISM LEVELS

Among other groups there is evidence of adaptive radiation. There are 55 species of millipedes, of which 24 are non-endemic. Some European genera are missing, but among the rest there are 25 endemic species of the genus *Cylindroiulus* which occupy a much wider range of niches than European species in this genus. Out of 46 terrestrial isopod species, 22 are endemic. Compared with Europe, there are again few genera, but a multiplicity of species, in some cases sister species being present on Madeira and Porto Santo. Seventeen of the endemic species occur on a single island only. The molluscs possess over 260 species (compared with 110 in Britain), of which 73% are endemic. Some of this diversity is a consequence of the fact that mountain and coastal areas provide very different habitats. In the mountains, there are species of snails with much reduced shells living in extremely damp mossy locations, while the coast has species which attach themselves tightly to rocks and can aestivate during dry conditions. Many islands have their own species closely related to others on different islands, so that it is evident that isolation between islands has promoted species formation. As with the millipedes and isopods, some European genera are lacking and there has been evolution of species in endemic genera to fill the niches they would have occupied. One or two taxa may represent examples of parapatric speciation, different mixtures of characters appearing in different populations so that no clear taxonomic distinctions can be made. These examples are, however, a small minority.

17.1.4 ABSENCE OF CONTINENTAL GENERA

In all groups, some of the characteristic mainland genera are lacking. The tendency to endemism is inversely associated with the dispersal powers of the group concerned. Where animals have high dispersive ability, as in the birds and butterflies mentioned above and also in beetles, repeated occasional introductions appear to take place, which must be matched by at least temporary extinctions. To some extent, therefore, the fauna is a consequence of a balance of accidental introduction and extinction. Gaps in the fauna are associated with broadening of the niches of species which are present. Subspecific, and more rarely, specific distinctness occurs, but there is no variation between islands.

17.1.5 SPECIES PROLIFERATION

In less mobile groups, the fauna differs between islands. In the molluscs, some species are common to several islands while others are restricted to a single island or islet. Some of the diversity, such as differences in the lowland snail faunas of eastern and western Madeira, may be the result of past volcanic events which have broken up the land into different portions from those we know now. There are fossil deposits containing extinct forms, indicating that species replacement has taken place while the fauna has evolved. The number of species occupying a single habitat on one part of an island is not unusually high, so that the great species richness is mostly the result of allopatric species formation on different isolated land masses, a process known as vicariance speciation. The point made by Wollaston, that several similar species in the same genus coexist on the same island, applies to at least six genera of beetles, and also to snails and isopods. This is likely to be the result of reinvasion, after the development of specific distinctness between the isolated parts of an ancestral species.

17.1.6 ROLE OF COMPETITION

Competition may have played differing roles in the evolutionary process in different groups. The broadening of niches in some bird species and in the lizard suggests that relaxation of competition is followed by adaptive response. Introductions should then result in niche constriction or replacements, and this may be occurring in the two butterfly species pairs which are changing their range. In the molluscs, 90% of the species are in four families, whereas in northwest Europe 10 families make up the same fraction. Local radiation has taken place, including the evolution of unusually small helicid species, which take the place of some of the absent small discoidal species. On the other hand, when one looks at the present-day distribution of non-endemic species of snails it seems that they simply add on to the endemic species instead of replacing them, so that sites with endemics and non-endemics have more species than those with endemics alone. Some collections which we made on Madeira and the Desertas had a mean of 4.6 species per site in samples with endemics only, and 6.5 in samples with non-endemics. The mean number of endemics in the latter samples was 4.9. This indicates either that competitive interactions are not particularly important in determining the faunal associations or that insufficient time has elapsed for competitive restriction to have taken effect in these samples.

Two opposed general views have been put forward as to the factors determining the species composition on islands. At one extreme, the fauna and flora are considered to be the consequence of a balance of accidental colonization and extinction, so that competition plays little part. Extinction is a function of island size and colonization is a function of size and nearness to the continental source of colonists, so that an indication of species abundance can be obtained from the size and position of the islands. This passive equilibrium theory was proposed by R.H. MacArthur and E.O. Wilson. At the other extreme, it may be argued that introductions are sufficiently common for the composition to reflect the number of niches, which, in turn, are evidence for competitive interaction. In Madeira there is some evidence for accidental introduction but many instances of local evolution, and some of these seem to provide evidence of competition. The significance of competition comes up again in another kind of floral and faunal comparison, that of the temperate versus the tropical regions.

17.2 COMPETITION AND COMMUNITY STRUCTURE

One factor which is important in relation to species abundance is climatic stability. The richest land snail faunas occur in leaf litter in equitable warm temperate forests on the North Island of New Zealand, which have been subject to minimal climatic disturbance. Likewise, the climate of Madeira is seasonally invariant and has been subject to little change over two million years compared, for example, with Britain, an island 300 times larger but with fewer than half as many snails. Does climatic stability lead to species abundance?

Writing in 1878 in *Tropical Nature and other Essays*, Alfred Russel Wallace answers this question as follows.

> A constant struggle against the vicissitudes and recurring severities of climate must always have restricted the range of effective animal variation in the temperate and frigid zones, and have checked all such developments of form and colour as were in the least degree injurious in themselves, or which coexisted with any constitutional incapacity to resist great changes of temperature or other unfavourable conditions. Such disadvantages were not experienced in the equatorial zone. The struggle for existence as against the forces of nature was there always less severe, – food was there more abundant and more regularly supplied, – shelter and concealment were at all times more easily obtained; and almost the only physical changes experienced, being dependent on cosmical or geological changes, were so slow, that variation and natural selection were always able to keep the teeming mass of organisms in nicely balanced harmony with the changing physical conditions. The equatorial zone, in short, exhibits to us the result of a comparatively continuous and unchecked development of organic forms; while in the temperate regions, there have been a series of periodical checks and extinctions of a more or less disastrous nature necessitating the commencement of the work of development in certain lines over and over again. In the one, evolution has had a fair chance; in the other it has had countless difficulties thrown in its way. The equatorial regions are then, as regards their past and present life history, a more ancient world than that represented by the temperate zones, a world in which the laws which have governed the progressive development of life have operated with comparatively little check for countless ages, and have resulted in those infinitely varied and beautiful forms – those wonderful eccentricities of structure, of function, and of instinct – that rich variety of colour, and that nicely balanced harmony of relations – which delight and astonish us in the animal productions of all tropical countries.

This is an interesting passage, because it conflates the two parallel strands in evolution, namely adaptation and speciation, as if to explain them both in the same terms. Environmental fluctuations are seen as throwing difficulties in the way of evolution, whereas we might expect them to promote species formation. Environmental constancy results in both a rich variety and a nice balance, whereas we would expect it to influence the balance but not necessarily the richness.

As one proceeds from the higher latitudes towards the equator, the annual and diurnal fluctuations in climate decrease. This aspect of climatic stability has always operated in the same way. The number of species increases and so does the amount of available solar energy. When numbers of individuals in the species of a community are counted, the relation of species number to abundance is found to be lognormally distributed. If the processes determining these empirical distributions are very powerful and common to all regions, then reduction in the total energy available will result in a progressive dropping out of the rarer species and the total number of species will go down. There is no obvious reason, however, why a greater amount of energy available to be converted to biomass must lead to a larger number of species, it could alternatively result in greater abundance.

The relative climatic stability does, however, have two predictable effects. The probability of extinction is likely to be reduced, and those species which occur are more likely to be operating near their equilibrium densities and to be K-selected, than in temperate regions with greater annual variation in climate. Because selection operates to increase the efficiency of intraspecific competition, they are likely to have lower rates of increase. This is consistent with Wallace's contention that species in equatorial regions often exhibit very precise adaptation to the niches in which they live. Nevertheless, the number of species present varies greatly between one equatorial region and another.

If we examine the number of butterfly species in various regions, for example, we see that there are many more in Europe than in Britain, and many more in Africa south of the Sahara than in Europe. The figure increases from 70 to 380 as one moves from Britain to Europe, and to 2600 in sub-Saharan Africa. In another tropical forested area, New Guinea, there are 500–600 species of butterflies, while South America has 8000–12 000 species. These great discrepancies are obviously to some extent a function of area. In Table 17.1 the raw data are converted to numbers per 100 000 square miles. On this scale, Africa has half as many species as Britain, much the same as Europe and very many fewer than South America or New Guinea. Table 17.1 also shows approximate figures for animals and plants in a number of other groups, which show that the discrepancy in species abundance between Africa and other regions containing tropical forest is not confined to butterflies. Some of these differences reflect adaptive radiations confined to a particular region. There are more ungulates in tropical Africa than in tropical America, more bats in America and similar numbers of species in the other mammal groups. Some groups of palms are African and Asian, others are exclusively American. Nevertheless, the discrepancies are so great as to show that, on average, Africa is very species poor in relation to comparable tropical regions. The difference would hold up if, as with the butterflies, numbers were to be converted to unit areas. Any explanation of the change in diversity from temperate to tropical regions has to be put forward in the light of these differences between comparable tropical biota.

One type of explanation for the differences concerns the climatic fluctuations over the past two million years, and their effect on distribution of habitat types. Glacial advances and regressions during the Pleistocene were synchronized with dry and wet phases in the tropics. The sea-level

Table 17.1 Numbers of plants and animals in different regions of the world. There is as great a discrepancy between different parts of the tropics as between them and temperate regions

(a) Numbers of butterflies. For areas the size of Britain these are larger than the number breeding at any one time, because climatic fluctuation affects breeding status while immigrants regularly arrive:

	Total species	Number per 10^5 miles^2
Britain	70	74
Europe	380	24
Ethiopian Africa	2600	31
South America	8000–12 000	130
New Guinea	500–600	165

(b) Numbers of plants and animals in particular groups in different parts of the tropics:

Species in flowering plant genera restricted to America and Africa	America 4019	Africa 573		
Species of palms	N. America 339	S. America 837	Africa 117	Eastern Tropics 1385
Genera of ants	New World 65	Africa 31	S.E. Asia 22	Australasia 32
Families and species of birds	Amazonia 49 (592)	Congo 37 (212)		
Parrot species	S. America 140	Africa 25	S.E. Asia 30	Australasia 150
Families and species of mammals	Neotropics 50 (810)	Africa 51 (765)		
of artiodactyls	3 (17)	5 (87)		

Data from various sources, including Meggers *et al.* (1973) and Keast (1969).

declined as water was bound up as ice, and increased during the warm phases. In South America, careful analysis of modern species distributions and of data from Pleistocene deposits, has led to the view that the vast forest area extending around the Amazon basin has at times been broken up into much smaller refugia, perhaps a dozen or more in number, during periods of dry climate. These periods of isolation of parts of previously continuous distributions resulted in allopatric species formation. While low rates of extinction and presence of K-selected species may be the result of annual climatic constancy, on this view the high species diversity derives from longer-term climatic fluctuation.

A difference in species diversity between regions is then interpretable if it correlates with the expected number of refugia. In the South East Asian area between Malaysia and New Guinea, sea-level changes have created archipelagos of islands which have been joined to a greater or lesser extent at different times in the past; here the refuges are a host of true islands and high species number would be expected. African rain forest falls into five components; upper and lower Guinea in west Africa, Cameroon, the Congo basin, and an eastern coastal fringe. It is probable that during dry phases, these forests contracted, but did not fractionate much further into numerous refugia. Diversity as a result of allopatric separation should therefore be relatively low, and that is the pattern observed. The consequence should be, however, that as progressively larger areas are examined, the rate of increase in number of species with area should be greater in regions with many refugia than in those with few. The theory suggests that the number of sympatric species in a given small territory should be the same, or if anything smaller, in regions with high rates of increase in species with area than in areas with low rates – the intercept of the curve should be equal or lower. This prediction is not usually met. In fact, greater overall species diversity tends to be associated with greater sympatric diversity. Examples are shown in Figure 17.1. In Figure 17.1(a) data for birds are presented, showing that figures for tropical African forest regions are nearer to those for North America than those for South America. One study in Central America recorded more resident bird species on a single estate than occur in the whole of the Congo basin. If it is reasonable to extrapolate, then a considerably lower sympatric diversity is indicated in tropical Africa than in tropical America. Figure 17.1(b) gives some values for butterflies which include samples of species caught in the same place. The regions of higher species diversity also have more species flying together.

If sympatric diversity increases when there is allopatric speciation, then more species coexist when more are formed. It appears to follow that species number in a given area is determined not by the number of niches, and the competitive interactions which they imply, but by the number of species which come into being. Perhaps competition is comparatively

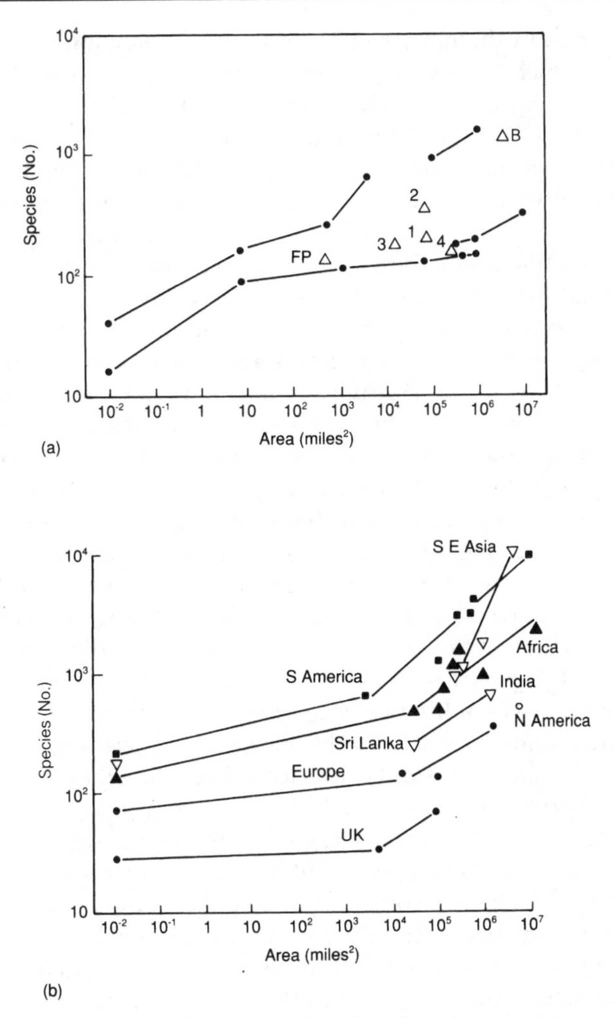

Fig. 17.1 Relation of number of species to area in different parts of the world. (a) Land birds. Solid points, lower line: North America. Solid points, upper line: South America. Triangles represent 1, Upper Guinea; 2, Lower Guinea; 3, Cameroon and Gabon; 4, Congo basin; FP, Fernando Po (Bioko); B, Brazil. Reproduced with the permission of The Linnean Society from *Biol. J. Linn. Soc.*, Cook (1978), data from various sources. (b) Butterflies. Reproduced with the permission of The Linnean Society from *Biol. J. Linn. Soc.*, Legg (1978).

unimportant in determining species number. This conclusion is also suggested when one asks what kind of competitive interaction there can be between assemblages of many species. The equations for competing pairs given in Chapter 4 describe the ways in which the competition can operate. Some patches of tropical forests have as many as 200 species of tree in them. Each individual of a given species is likely to be some distance from the next, so that each is surrounded by individuals of many different species. Although the net effect of these others may inhibit growth and reproductive success, it is hard to see how there can be distinct competitive pairwise interactions between them all. If we take the multispecies equation from section 4.3, then for a forest of k similar tree species, R_i and K_i must have not dissimilar values. If the trees have come together as a result of evolutionary processes which are random with respect to each other, then the a values for i not equal to j might be randomly distributed. It is easy to see that under these circumstances the mean value of a must be inversely related to the number of species if they are to coexist. If there are 200-odd species, then for any one of them an individual of each of the others can have no more than a negligible effect compared with an additional individual of its own species. Investigation of models involving random a matrices has shown that the number of sustainable species rapidly reaches a ceiling, and also that it is possible to end up with disconnected small groups of interacting species. The implication is that in species rich communities, other species may be a component of the environment which has an inhibiting effect *en bloc*, but they do not normally exhibit precise multiple pairwise interactions. Success depends on whether the environment is one in which a species can survive and achieve an equilibrium density, but with few exceptions, is independent of which particular species are already present.

The situation where the presence of particular species is important is where subdivision of range has led to vicariance speciation, and the newly distinct species are very similar to each other when they come together. This could happen repeatedly when there are geological catastrophes such as volcanic eruptions on islands or large fluctuations in climate over periods of the order of one or two million years. Then the new species will interfere directly with each other. The most likely outcome is extinction of one or other, and they will only coexist if there is niche divergence, or clumping and patchiness of distribution, or some other adaptation which reduces the size of the competitive effects. This is probably the explanation of the guilds of species which are frequently found.

17.3 CONCLUSION

Some of the most important problems in current ecological research concern the patterns and causes of species diversity. The intention here is to present a framework within which these problems can be viewed, and in that spirit we come to the following conclusions. Climatic and ecological instability promote speciation. Fractionation and recontact will result in the formation of guilds of similar species, which have become adapted so that they are not in direct competition. In their case, competition will rarely be detected, but has been an essential factor in generating the pattern seen. At a more distant level of relatedness, interspecific competition is not of predominant importance in determining whether or not a species is successful; it is part of the totality of biotic and physical factors to which the species has to adapt. In regions where annual climatic variation is low, and where there is climatic stability over historical periods of a century or more, species tend to reach equilibrium densities and to be K-selected. Where there is greater short-term variability, they are more likely to be r-selected. The probability of extinction is also reduced in climatically stable conditions. The equilibrium diversity is determined by the balance of physical factors promoting speciation and those regulating extinction, so that greatest diversity occurs in regions which are climatically stable over years or hundreds of years, but are geologically or climatically unstable on a time scale of 10^5 or 10^6 years. In the high Arctic, solar energy is severely restricted for half the year; abyssal faunas depend for their energy source on organic material raining down from shallower water. In limiting cases such as these, total energy available in a region, and the way it is spread through the seasons, must restrict the number of species which can coexist. Energy availability probably has little to do, however, with the differences in species abundance between terrestrial floras and faunas of temperate regions and of the tropics.

There are parallels between explanations of genetic variability and those relating to species diversity. Allopatric speciation results from random fractionation of the geographical range of species, followed by development of isolating mechanisms. DNA sequence divergence may assist this process by promoting isolation, and it provides a measure of the timing of speciation events. When recently separated taxa make contact again, there may be replacement of one by another or adaptation to remove competition and produce niche specialization. Coexistence is then possible, rather as multiniche polymorphisms are maintained. Nevertheless, communities are largely composed of assemblages of species without tight networks of competitive interactions. At the genetic level, the analogy would be with functionally equivalent alleles at loci influencing a biometrical character; for each there is an expectation of persistence in the population which depends on the dynamics of introduction and loss, but they are independent of each other. When we consider evolution of a fauna such as that of Madeira, however, the good stories concern adaptation, a product of natural selection, while variation in the neutral component of the genome is not apparent. The stories may not be agreed. A case interpreted as parapatric speciation by one researcher may seem to be allopatric speciation to another. Population genetic theory does, however, provide a satisfactory framework within which evolutionary problems may be discussed.

One of the most important areas for future research is the study of limits to adaptation. What is, and what is not, possible in the way of response to selection depends on developmental and coadaptational constraints.

These constraints may be studied by examining such things as the genetics of pattern formation in embryogenesis, the effect on development of variation in the timing of gene action, and the effect of mutational order on the phenotype which arises in a particular lineage. When we know more about them we shall have a better understanding of biological diversity. The process of working from the genes up is sometimes criticized on the grounds that the phenotype cannot be uniquely predicted by knowing the genes, the future evolution of a group by knowing the selective forces to which it is subject. As we move upwards through the hierarchy of levels of complexity, emergent properties become apparent which cannot be anticipated at a lower level. That does not mean, however, that one should not try to understand how simpler components unfold into complex systems. It is a fact of life that genetic information passes from one individual to another as little more than a strand of DNA. The excitement lies in getting some idea of how it is transformed from this elementary state into the world of diversity, beauty and grandeur which we experience.

17.4 SUMMARY

In trying to explain the proliferation of species on Earth we need to examine a series of possible causative agents. These include the following: the trophic system, habitat variation, competitive interactions, the mode and causes of speciation and the conditions leading to extinction.

Species diversity is to some extent the result of the trophic system, the variety of habitats and the existence of niches within habitats. The concept of the niche entails the idea of interspecies competition. The differences between trophic levels are trivial in explaining species richness. There are reasons to believe that species–species interactions are not important in accounting for communities of many species, and to that extent the concept of the niche is not particularly helpful in explaining diversity. The number of habitats is clearly of central concern – the more there are the more species may exist. The difficulty is that we do not know how adaptive constraint limits the habitat range which can be tolerated by a given species. If that were known the problem would be all but solved.

A variety of evidence, among it the evidence of adaptive radiations on isolated islands and continents and in isolated bodies of water, indicates that allopatric speciation is the most common form in the most species rich groups of organisms. Species are therefore generated most rapidly in regions which are geologically unstable over time scales of a few hundred thousand to a few million years. Annual climatic stability and climatic patterns which are stable over hundreds of years reduce the chance of extinction. The climatic and geological stability in a given region therefore generate a balance of speciation and extinction which broadly determines the number of species present. What we should do next in the quest for

explanations is to examine critically the ways in which adaptive constraint may be related to ecological tolerance.

17.5 FURTHER READING

Diamond, J. and Case, T.J. (eds) (1986) *Community Ecology.*
Grant, P.R. (ed.) (1997) *Evolution on Islands.*
May, R.M. (1973) *Stability and Complexity in Model Ecosystems.*
May, R.M. (ed.) (1981) *Theoretical Ecology.*
Williamson, M. (1981) *Island Populations.*

Pattern and process – the long-term evolutionary perspective 18

18.1 INTRODUCTION

A more extended perspective is required to discuss development of communities than of genetic diversity within species, although in both cases explanations involve the parallel operation of random events and natural selection. When we consider genetic systems themselves a still longer view is indicated. Breeding systems have so far entered the scene as if they were the given plan which allows evolution to take place. In fact, they arose as genetic innovations themselves, which may be discussed in terms of short-term evolutionary processes.

If a mutant gene arose in a sexually reproducing species which allowed a female to produce all-female offspring parthenogenetically, then any female carrying it would have twice the reproductive capacity of a sexually reproducing female. There is therefore very strong selection for the sexual system to disappear. Three types of reason have been suggested for its development and almost universal prevalence.

1. Cell cultures, and protozoan cell lines which are not permitted to conjugate, have a limited expectation of life. A reasonable explanation is that mutants accumulate at mitosis, causing a progressive decline in fitness. Recombination would allow deleterious mutants to be eliminated. It is possible that at the very beginning of the evolutionary story recombination mechanisms arose for this purpose.

   The problem is a continuous one in asexual populations. As deleterious mutants accumulate, a population of limited size develops a distribution of lineages with 0, 1, 2, 3, etc. mutants with correspondingly declining fitnesses. Individuals of the most successful lineage, with no mutations, are likely to be quite rare and risk being lost by chance, while the accumulation of deleterious mutants is continuous. H.J. Muller likened the result to the operation of a rachet, which is constantly tending to increase genetic load while selection opposes the trend by eliminating the most unfit, and his term has become established in the literature. Genetic recombination minimizes this effect.

2. Some mutants are advantageous, however. In an asexual species, a clone of individuals containing an advantageous mutant increases in frequency at the expense of other clones, but incorporation of a second advantageous mutant requires a second mutation. In a sexual species, a second mutant from a different lineage is rapidly incorporated through genetic recombination and the process of adaptation is greatly speeded up.
3. The advantage may occasionally be more direct. In a fairly constant environment, the optimal phenotype will be selected and recombinants will be, on average, disadvantageous. When environmental change is rapid enough, however, recombinants could provide the necessary new phenotype for survival.

Systems which ensure that there are two sexes to exchange gametes are often polymorphisms. In most animals, there is an XY system which basically consists of one or a few genes, although often a whole chromosome becomes differentiated. Sex is not always determined in this way. Some molluscs and fish start life as males and later turn into females. In some reptiles, sex depends on incubation temperature. Many animals and plants are hermaphrodites, and sometimes but not always they have mechanisms which prevent self-fertilization. One of the distinctive features of these genetic mechanisms regulating breeding behaviour is that each is usually characteristic of a large taxonomic group. Thus, particular families of fungi and flowering plants are likely to contain species with self-incompatibility systems. In animals, some orders of insects and phyla of vertebrates have heterogametic males (with sexually differentiated gametes) while in others the females are heterogametic. Examples of this kind of constancy within major taxonomic groups not only illustrate the degree to which the various systems have been conserved but also highlight evolutionary trends in plants and animals.

18.2 SELF-INCOMPATIBILITY IN FLOWERING PLANTS AND FUNGI

In the fungi, some members of the taxonomic family Phycomycetes such as *Allomyces arbuscula* produce only one mating type: they are homothallic. Such species regularly show haploid selfing, leading to the immediate and total elimination of both homozygosity and recombination. By contrast, the majority of fungi have self-incompatibility systems which prevent haploid selfing and restrict diploid selfing. Because these systems involve separate mating types, the fungi are said to be heterothallic. In taxonomic families such as the Phycomycetes, including the bread mould *Rhizopus nigricans*, and the Ascomycetes, including the bread yeast *Saccharomyces cerevisiae*, self-incompatibility is controlled by a single locus of two alleles. Each zygote is therefore heterozygous for the mating-type gene. Only half the possible matings between the products of a single meiosis will be compatible, with the result that the frequency of diploid selfing is restricted to 50%.

The family Basidiomycetes consists of the most sophisticated fungi, including the heterobasidiomycetes (rusts and smuts) and the eubasidiomycetes (brackets, mushrooms and toadstools). The life cycle is complex, normally including a monokaryotic phase consisting of hyphae with only one type of haploid nucleus and a dikaryotic phase formed by the fusion of two monokaryons of opposite mating type. The cells of the dikaryons each contain two haploid nuclei of opposite mating type. In this way, the cell is effectively diploid although the nuclei remain haploid. The two types of nuclei only fuse to form a diploid nucleus in the basidium. The diploid nucleus immediately undergoes meiosis to form four haploid products which enter the basidiospores. Dikaryons have proved of great value in genetical research. Non-allelic mutants will complement each other in a dikaryon, restoring the wild-type phenotype, while allelic mutants will not. Dikaryons have a distinctive mode of cell division which ensures that each new cell receives one haploid nucleus of each type. The cells are separated along the hypha by septa each of which is perforated by a specialized pore. When the cell divides, an additional bridging tube is formed between the old and new cells, which by-passes the septum. This bridging tube is known as a clamp connection. Once the nuclei have divided, one daughter nucleus passes through the pore to the new cell and a daughter from the other division passes around the clamp connection.

The main incompatibility system in basidiomycetes stems from the distinctive mode of cell division. In some members of the eubasidiomycetes, such as the field mushroom *Agaricus campestris*, the lawyer's wig *Coprinus cinereus* and the split-gill *Schizophyllum commune*, self-incompatibility is controlled by two unlinked loci (A and B) each of two alleles. Monokaryons with A alleles in common will fuse to form dikaryons lacking a pore whereas dikaryons sharing a B allele will lack clamp connections. Neither type is able to grow and develop reproductive structures. Since compatible monokaryons must differ at both loci, all diploid nuclei are heterozygous for both mating-type genes. The four products of a single meiosis will have distinct mating types: AB, Ab, aB and ab. If these are combined in all possible pairwise combinations, only four of the 16 combinations will be compatible, thereby, restricting the frequency of diploid selfing to 25%.

The common ink cap *Coprinus comatus* has an unusual self-incompatibility system, consisting of a single multi-allelic mating-type gene. Like the system in the Ascomycetes, this leads to 50% diploid self-compatibility. The highly polymorphic nature of the mating-type gene results in a high level of general compatibility within the populations and is indicative of a high frequency of outcrossing. The cultivated mushroom *Agaricus bisporus*, by contrast, shows a pattern of reproduction akin to parthenogenesis. Only two spores are produced following meiosis in the basidium, each containing two haploid nuclei of opposite mating type so that every spore has the capacity to produce a dikaryon on germination. Mating between separate monokaryons is avoided.

Evidently, the different families of fungi illustrate a trend towards a reduction in the frequency of diploid selfing. Increasing numbers of mating-type loci could successively reduce the degree of diploid self-compatibility by 50% for each new locus but such systems would become ever more unwieldy. While the frequency of diploid selfing could be greatly reduced by such means, it could never be reduced to zero because the incompatibility reaction in fungi is always between two gametophytic tissues.

In all non-flowering plants, mating recognition involves only gametophytic cells and, so far as we are aware, only molecules of gametophytic origin. Haploid selfing is prevented either (1) by separation of gametophytes into different mating types (heterothallism) or sexes (haplodioecy) or (2) by polyploidy so that the gametophytes are no longer haploid. Examples of separate gametophytes are self-incompatible fungi, haplodioecious bryophytes (e.g. the moss *Mnium undulatum*), heterosporous pteridophytes (e.g. the lesser clubmoss *Selaginella kraussiana*, $n = 10$) and all gymnosperms (conifers and cycads). Polyploidy is common in homosporous clubmosses (e.g. the fir clubmoss *Lycopodium selago*, $n = 132$) and ferns (e.g. the adder's tongue fern *Ophioglossum vulgatum*, $n = 240$); here, though gametophytes may self-fertilize, they are not haploid, so any heterozyosity within the gametophyte persists in the sporophyte. None of these mechanisms can prevent sporophytic selfing although polyploidy mitigates its effect by slowing down the loss of heterozygosity. Some degree of selfing therefore seems to be inescapable in diploid non-flowering plants, unless they exhibit separate male and female sporophytes, a condition termed diplodioecy. Many gymnosperms are diplodioecious, notably the maiden-hair tree *Gingko biloba* and the yew *Taxus baccata*.

Flowering plants first appeared in the geological record at the beginning of the Cretaceous period ($125–145 \times 10^6$ years BP), some 100 million years after the first mammals which are known from late Triassic deposits ($205–225 \times 10^6$ years BP). Their mating behaviour exhibits two novel advantageous features: double fertilization and the interposition of maternal receptive tissue between pollen and egg, the latter feature causing us to refer to flowering plants as angiosperms (enclosed seeds). Double fertilization produces a unique type of nurse tissue – endosperm –

which is intermediate in genotype between zygote and nucellus and which is commonly triploid but may be tetraploid in some groups of species.

The other unique feature of angiosperms is the enclosure of the ovule which forms the seed within an ovary which eventually becomes the fruit. Pollen is intercepted first by the stigma and then by the style. Self-incompatible flowering plants thus have the great advantage that the incompatibility reaction is not between two gametophytic tissues, as in lower plants, but between male gametophytes and female sporophytes. Diploid selfing can be avoided because all the information about the compatibility of the diploid parent is contained within either the stigma or the style.

A greater range of breeding systems is found in angiosperms than in all other plants. While all flowering plants have unisexual gametophytes (haplodioecy), some such as the European hop *Humulus lupulus* and the red campion *Silene dioica* have unisexual sporophytes (diplodioecy). Some species have breeding systems which discourage selfing by preventing it from taking place within a flower (autogamy) yet allowing occasional selfing between flowers (geitonogamy). Such systems include monoecy where there are unisexual flowers on the same sporophyte, for example in maize (*Zea mays*), protandry where the stamens mature before the stigmas, as in the foxglove *Digitalis purpurea*, and protogyny where the stigmas mature before the stamens, as in the ribwort plantain *Plantago lanceolata*. Some of these breeding systems are characteristic of the taxonomic family in which they occur, for example protandry is commonly found in the mallows (Malvaceae) and saxifrages (Saxifragaceae). The most distinctive examples are provided by the wide range of self-incompatibility systems. As in the fungi, these systems illustrate major evolutionary patterns.

In contrast with those in fungi, all self-incompatibility systems in flowering plants eliminate or virtually eliminate diploid selfing. They differ in genetical and physiological control, in morphology and, most importantly, in their ability to restrict mating of siblings (brother–sister mating). The most sophisticated systems can place greater restrictions on inbreeding than does diploid sexual dimorphism, not only by comparison with dioecious plants such as the red campion *Silene dioica* but also in comparison with most animals. The efficiency with which it is restricted can be estimated as the ratio of sib / general compatibility, where general compatibility represents the likelihood of two randomly selected individuals from the population at large proving to be compatible.

The most obvious self-incompatibility systems are those where the mating types have distinctive morphologies. These heteromorphic systems are of two types: the distylic found in the Primulaceae (primrose family) and the tristylic found in the Lythraceae (loosestrife family),

Oxalidaceae (sorrel family) and the Amaryllidaceae (amaryllis or daffodil family). In the primrose, *Primula vulgaris*, the petals (corolla) are fused at the base to form a tube. The stamens form a ring within the tube. One mating type has a long style with a stigma at the entrance to the corolla tube, giving the flower a pin-eyed appearance. The stamens are out of sight, halfway down the corolla tube. The other mating type has a short style with a stigma halfway up the corolla tube. The stamens are clearly visible at the top of the tube; they resemble a cut thread which is known as a thrum in the textile industry. The flowers are therefore thrum-eyed.

Both mating types set small amounts of seed on selfing. Pin breeds true whereas thrum segregates 3 thrum:1 pin. Thus pin seems to be homozygous for the recessive form of the self-incompatibility gene (ss) whereas thrum is heterozygous (Ss). In fact, at least six genes are tightly linked within a self-incompatibility super-gene. Genes affecting female characteristics of the flower – style length, length of stigmatic papillae and stigmatic compatibility – are in one region of the super-gene while those affecting male characteristics – height of stamens, pollen size and pollen compatibility – are in another section of the super-gene.

Recombination within the super-gene produces self-compatible recombinants known as homostyles. Short homostyles with female thrum characteristics and male pin characteristics have been seen in experimental breeding programmes but do not succeed in natural populations. Long homostyles with female pin characters and male thrum characters occur naturally at a frequency of about one in a thousand. Not only are they self-compatible but, because the stigma is surrounded by stamens, they almost invariably self-pollinate. A simple recombination event has caused the whole self-incompatibility system to break down. Some populations of primroses in southern England have very high frequencies of long homostyles (up to 80%).

While the distylic system in primroses more or less eliminates diploid selfing, it is completely ineffective at eliminating sib-mating. This is because half the offspring of a compatible mating will be pin and half thrum. This 50:50 ratio is the same as that of the two mating types in the population at large. Siblings are thus at no greater disadvantage in mating than two individuals chosen at random from the general population and the sib/general compatibility is unity.

The incompatibility reaction in primroses takes place at the stigma. Whereas compatible pollen germinates and penetrates one of the stigmatic papillae, the entry of incompatible pollen is blocked by secretions of an inhibitory substance (callose), not only by the papilla but also by the pollen itself. It is an example of mutual rejection. The incompatibility reaction is determined by antigenic proteins embedded within the outer wall of the pollen grain. These proteins come not from the pollen itself but from the lining of the anther wall known as the tapetum. The self-incompatibility system is therefore sporophytic in the sense that the self-incompatibility reaction of the pollen is determined by the genotype of the male sporophyte (tapetum) rather than the male gametophyte (pollen).

Dominance and sporophytic control are characteristic of all heteromorphic self-incompatibility systems. Tristylic self-incompatibility systems are characterized by three different mating types: short-styled, medium-styled and long-styled. There are two diallelic loci: S/s and M/m. S/s is epistatic to M/m. Thus all genotypes carrying S are short-styled, while those lacking S but carrying M are mid-styled. The long-styled type is a double homozygote for the recessive genes (s and m). Examples of tristylic systems are provided by the diploid *Oxalis valdiviensis* and the purple loosestrife *Lythrum salicaria* which is tetraploid. Most species of *Narcissus* are tristylic and these represent a wide range of ploidy from diploid to hexaploid. Tristylic systems are only slightly more efficient at restricting sib-mating than distylic systems, average sib/general compatibility being of the order of 90%.

The most efficient systems of self-incompatibility have numerous mating types which are indistinguishable morphologically; they are homomorphic. These systems all have multi-allelic self-incompatibility genes. Some exhibit gametophytic control with the compatibility of the pollen being determined by antigenic proteins synthesized by the pollen grain itself. Here the self-incompatibility reaction takes place, not at the stigma, but in the style. Again, callose is secreted but in these systems it surrounds incompatible pollen tubes, depriving them of nourishment. Because the control is gametophytic, there is no dominance. Examples of gametophytic self-incompatibility are found in the Onagraceae (willow herb family), for example in the evening primrose *Oenothera organensis*, in the Rosaceae (rose family), for example in top fruits such as the sweet cherry *Prunus avium* and self-incompatible varieties of apples (*Malus sylvestris*) and pears (*Pyrus malus*), and in the Fabaceae (pea family) where more than 450 s alleles have been identified in the red clover *Trifolium pratense*, as many as 45 being found in a single population. Where large numbers of s alleles are present in a population, the sib/general compatibility is close to 75%. This represents a marked improvement over the heterostylic systems but it is still fairly ineffective at restricting sib-mating. Studies of *Oenothera organensis* have shown that self-incompatibility breaks down in autopolyploids. The presence of two self-incompatibility genes within the same pollen grain produces a mixture of antigenic proteins which confounds the detection system in the style.

In terms of population genetics, these self-incompatibility alleles generate a frequency-dependent polymorphism. Only heterozygotes can exist because no pollen tube develops in a style carrying the same allele as the pollen. The minimum number of alleles is therefore three, and it is not known how the system originally comes into being. Once there are three alleles, there is an opportunity for further accumulation. For k alleles, there is a stable equilibrium when all are at a frequency of $1/k$. Any new mutant has the advantage of rarity and all alleles will tend towards the new lower equilibrium. There is therefore an allelic competition reason for the number of alleles to increase, apart from any benefit which may arise from the restriction of inbreeding. The advantage of rarity lessens progressively as k increases. A balance will therefore eventually be achieved which depends on drift and on other selective considerations.

Self-incompatible grasses (Poaceae) also exhibit homomorphic gametophytic control but two multi-allelic self-incompatibility genes are present: s and z. The z gene is thought to have been derived from the s, possibly by duplication. All matings are compatible unless pollen and style share alleles at both loci. This system does not break down in polyploids but it is less efficient than the single-locus system at restricting sib-mating. The average sib/general compatibility is likely to be of the order of 90%.

Efficient self-incompatibility systems have homomorphic mating types but sporophytic control with the incompatibility reaction taking place at the stigma, as in the primrose. There is a single multi-allelic self-incompatibility gene but a complicated system of dominance. Some alleles are dominant in the pollen, the stigma or both. This system of multi-allelic, sporophytic control is found in the most sophisticated family of flowering plants: the Asteraceae (daisy family), for example in *Cosmos bipinnatus* where it was first discovered. It is also found in the Brassicaceae (cabbage or mustard family), where it is of considerable commercial importance. Studies of *Brassica oleracea* (cabbage, brussels sprout, etc.) indicate that about 63% of s alleles show codominance so both alleles in a diploid are always expressed in the pollen and in the style. Two parents heterozygous for different codominant s alleles will be fully compatible and produce four different mating types in their offspring. The offspring will only be compatible if they have no s allele in common. Under these circumstances, sib compatibility is reduced to 25% while the general compatibility in such highly polymorphic systems is likely to be close to 100%. The sib/general compatibility is therefore likely to be about 25%, representing a considerable limitation on sib-mating.

In surveying the restrictions on inbreeding in plants and fungi, we see a progressive succession of genetic controls. In the lower plants and fungi, haploid selfing can be eliminated, either by self-incompatibility or polyploidy. In diploids, diploid selfing can be restricted but not eliminated because the incompatibility reaction is always between two haploid tissues. In polyploids, haploid selfing is not possible by definition although gametophytic selfing will be common. The evolution of the style and stigma in the angiosperms allows diploid selfing to be fully prevented. In the angiosperms, self-incompatibility systems differ not in their ability to control diploid selfing but rather in the degree to which they limit sib-mating. Within this spectrum, taxonomic families of flowering plants exhibit a wide range of self-incompatibility systems. Some are heteromorphic while others are homomorphic. Some display sporophytic control and dominance; others display gametophytic control and no dominance. Each system seems to be characteristic of all self-incompatible species within the family and hence to have been strongly conserved.

18.3 CONSERVATION OF MATING SYSTEM IN MOLLUSCS

Mollusca comprise the most species rich phylum of organisms, after the insects. The four less numerous classes in the phylum, including the cephalopods, are exclusively marine and all have separate sexes. Most bivalves are marine and about 90% of species have separate sexes. The class Gastropoda, which includes all those we think of as snails, is more diverse. It includes the subclass Prosobranchia, which are almost all marine and almost all (97%) with separate sexes. The two other subclasses are the Opisthobranchia, including the sea slugs, which are marine and almost entirely hermaphrodite, and the Pulmonata, the freshwater and land snails. These, too, are hermaphrodite. Why do we get such different strategies in different groups? In order to explore this question, we need to consider the advantages or drawbacks of hermaphroditism. This question is also tied up with the advantages and drawbacks of cross- and self-fertilization.

There are three types of reason why species should be cross-fertilizing hermaphrodites, and some probable costs associated with the system. Reasons for hermaphroditism are (1) that individuals may have difficulty during their lifetimes in finding individuals with which to mate, (2) that there may be an advantage in spreading the reproductive effort over a long period of life and accompanying change in body size, and (3) that if offspring live very close to their parents throughout their lives, there is an advantage in investing in both kinds of gametes, because excess effort in either direction is likely to lead to diminishing returns. As the number of male gametes goes up, there will come a point when no more female gametes are within reach with which they can unite. As the number of female gametes increases, so will the potential for sib–sib competition. A distribution of resource through both routes is therefore favoured; this is a non-obvious consequence of low mobility.

Explanation (1) may be invoked for the many cases in different animal groups where some species are both parasitic and hermaphrodite. Parasites combine a low probability of meeting a mate with a physical environment which is often more than usually stable, even though they may have to contend with evolving defence mechanisms on the part of the host. The relative constancy of the environment reduces the need for recombination and offsets the drawbacks associated with the more complex structural organization. This pattern is seen in its fullest development in internal parasites such as the tapeworms, which are little more than repeated units of self-fertilizing hermaphrodite genitalia. Among different groups of Crustacea, a route can be seen to this condition. Most are bisexual with a genetic switch mechanism, but there are exceptions among sessile forms, including parasites, and in species which require a resistant over-wintering stage. In the Cladocera (water fleas), seasonal change occurs from sexual reproduction to parthenogenesis. The barnacles are almost all hermaphroditic (although cross-fertilizing). Their relatives, the Rhizocephala, are self-fertilizing parasites. Some isopods are parasites on other crustacea, and have a facultative sex-determining mechanism that depends on the presence or absence of another parasite already established on the host when the new one arrives. If the host is unparasitized, the settling individual develops into a female, but if a female is already present the second parasite becomes a male.

It is unlikely that explanation (1) is important in pulmonate snails, since although, or perhaps because, they are immobile they generally live in patches of relatively high density. They are often also long-lived, so that the chance of passing a lifetime without finding a suitable mate is small. The second explanation can be imagined in some circumstances, for example when snails live in relatively severe Mediterranean climates. It may then be advantageous to exchange sperm at one time of the year, and having done so, to use them to fertilize eggs later, perhaps after accumulation of substantially more energy during a period when more food is available. This pattern of reproduction can be seen in helicid snails, which, although now more widespread, probably originated under such conditions. Eggs and sperm develop side by side separated by a layer of epithelium that covers the eggs while sperm are free to be released. Later the remaining male gametes degenerate, the separating layer breaks down, and the eggs can be fertilized. Copulation between individuals is reciprocal and commences early in the season; the received sperm are stored until fertilization takes place.

Most species probably live in climates which are not so restrictive. All are relatively immobile, however, and have to lay their eggs where they live as adults. Offspring–parent and sib–sib competition are therefore possibilities, and hermaphroditism may reduce the likelihood that they occur. Earthworms are another group with similar traits, to which the same argument may be applied.

Hermaphroditism offers the opportunity for self-fertilization. Whereas most pulmonates are cross-fertilizers, some very successful species self. One well-known example, *Rumina decollata*, is an adventitious colonizer of disturbed waste ground which often reaches very high densities where it is established. Among the slugs, species which live in rich environments generally cross-fertilize while selfers colonize more extreme habitats. These examples support the adaptive explanations discussed earlier with respect to plants, although a comprehensive survey would be required to decide whether the association generally holds true.

The land-living pulmonates differ in an important respect from prosobranch molluscs. Whereas adult prosobranchs may be as restricted in their movement as are adult pulmonates, the group is basically marine, rather than terrestrial, and the reproductive strategy involves separate sexes and a planktonic larva. Many exceptions occur, some of which can be explained as responses to particular conditions. For example, some slipper limpets (*Crepidula* spp.) are hermaphrodites with planktonic larvae which change from male to female as they develop. Because they settle upon each other, it is easy to see that there is an advantage to the newly settling individual being the right sex to mate with one which is already established. Periwinkles (*Littorina* spp.) have separate sexes and some have planktonic larvae while others brood eggs. An explanation of the difference between species may be sought in terms of differing need for local adaptation. Such examples are exceptions, however, rather than the rule. The existence of a planktonic dispersal phase removes the direct competition effect and, arguably, the selective pressure to be hermaphroditic.

We can thus see many cases where the reproductive pattern fits with the pattern of life, as if natural selection actively tailors a flexible system. Nevertheless, the Mollusca also raise a problem, pointed out by G.C. Williams (1975) and emphasized by Heller (1993), namely, that whole groups at a high taxonomic level have a fixed pattern of reproduction. Prosobranchs and pulmonates have different patterns of life, marine versus terrestrial, which may make it appropriate for one group to have separate sexes while the other is hermaphrodite. Pulmonate reproductive behaviour is very constant, however, while prosobranchs show some variation. But most importantly, the opisthobranchs, which are taxonomically close to the pulmonates, are hermaphrodite but have the habits of life of the prosobranchs. Does some kind of phylogenetic constraint operate, above the level of ecological adaptation, which renders some reproductive patterns almost universal within major groups? In the present case, the answer may lie with the consequences of simultaneous cross-fertilizing hermaphroditism, discussed in detail by Janet Leonard (1990). We have argued here that trade-off considerations favour hermaphroditism in molluscs with restricted powers of dispersal. In circumstances affecting some species, however, adoption of one role in mating may be more rewarding than adopting the other. If there are many individuals around, for example, then a hermaphrodite could gain more by acting as a male than as a female, because it could inseminate a succession of individuals and avoid the metabolic cost, time and possible danger of laying eggs. In principle these species should again evolve separate sexes. However, the immediate selective pressure now acts between pairs of reciprocally mating individuals. The pressure to cheat is counteracted by selection in the cheated to avoid suffering the consequent reduction in output. This may be why courtship patterns and sexual organs are often highly complex in pulmonates. If a hermaphrodite strategy has come to be well embedded, then reciprocal selection of this kind may inhibit reversion and fix true reciprocity. Once established, hermaphroditism becomes difficult to lose.

18.4 EVOLUTION OF SEX CHROMOSOMES

Genetic control of sexual dimorphism does not always give rise to morphologically distinct sex chromosomes. It has not done so, for example, in fish such as the guppies (*Lebistes* spp.) and flowering plants such as dog's mercury (*Mercurialis* spp.). In the guppy, males are heterozygous for the sex-determining gene and hence heterogametic while females are homozygous and homogametic.

Where several genes arise with the ability to influence sex determination, natural selection will favour divergence of distinct male and female combinations. This situation constitutes an example of linkage disequilibrium, where certain favoured combinations of genes are more frequent than the products of the individual frequencies would suggest. Because the accumulation of unisexual effects is threatened by recombination, selection will favour their clustering on a single chromosome. Eventually, selection pressures against recombination within the sex-determining region will lead to sufficient divergence for the male and female sex-determining regions to become non-homologous. As a result, morphologically distinct (heteromorphic) chromosomes become established, distinguished by non-homologous differential segments containing sex-determining genes and homologous pairing segments which ensure the regular co-orientation and segregation of sex-determining factors. The differential segment may be small relative to the pairing segment, as in the Hawaiian fly *Telmatogeton abnormis* or the mosquito *Aedes aegypti*, or it may be so extensive that only a minute pairing segment remains, as in the tsetse flies (*Glossina* spp.). Extensive differential segments usually contain large regions which are non-coding, for example in the sorrel *Rumex acetosa*, in tsetse flies and in man. The gradual transition from undifferentiated homomorphic sex chromosomes to a markedly heteromorphic pair is well illustrated by comparisons of species of snakes (Ophidia) and ratite birds (emus and rheas).

One of the important consequences of recombination is separation of advantageous and deleterious genes so that the latter are eventually removed from the population by selection. As we saw in *Oenothera*, chromosome segments which do not engage in recombination steadily accumulate deleterious or lethal mutants. The differential segment on the Y chromosome is as genetically isolated as the interchange multiples in *Oenothera* but it is not balanced by an equivalent segment on the X chromosome, so that lethal genes cannot become established. Nevertheless, mildly deleterious mutants will accumulate because they cannot be removed by recombination. Over the course of time, differential segments which are free of deleterious mutants will become progressively rarer, partly because of the relentless slow march of mutation and partly because some will be lost by chance. Loss due to genetic drift becomes progressively more important as the mutant-free segments become rarer. The proportion of segments with successive numbers of mutants is raised every time a mutant-free segment is lost (Muller's ratchet mechanism).

The effects are off-set by inactivation of the deleterious genetic material, often via methylation of the DNA, and by the accumulation of highly repetitive DNA sequences. A special class of repetitive sequences has been identified with the differential segments of sex chromosomes. These are interspersed amongst the differential segments of snakes but highly localized in those of mammals such as the mouse. The inactivated region remains permanently condensed and is known as constitutive heterochromatin, in contrast to the normal chromatin which is diffuse during nuclear interphase and known as euchromatin. Extensive differential segments usually contain large amounts of constitutive heterochromatin, for example in the sorrel *Rumex acetosa*, in tsetse flies and in man. Such genically inactive chromatin not only bears testimony to the neutralization of erstwhile deleterious mutants which might have accumulated during the evolution of the differential segment but also serves to accentuate the non-homology and genetic isolation of the differential segment.

Sex chromosomes exhibit an extensive range of sizes relative to the rest of the chromosome complement (autosomes). This is especially clear in beetles where species in families such as the Cassididae have sex chromosomes which are half the size of the autosomes while some members of the Alticidae have giant X and Y chromosomes which are many times larger than the largest autosome. The most extreme example is provided by the alticine species *Walterianella venusta* where the two sex chromosomes are 40–70 times the length of the largest autosome and, despite the presence of 44 autosomes, represent 70% of the genome in males ($2n = 44 + XY$). A high proportion of the genotype is therefore likely to display sex-linked inheritance. The giant sex chromosomes of male alticine beetles are also unusual in that they are achiasmate and attached to a separate spindle from that of the autosomes at the first division of meiosis. This situation suggests that the differential segment extends over the whole of the Y chromosome. Unfortunately, females do not appear to have been studied in any of these species so that we are not able to distinguish which sex chromosome is the X and which the Y (White, 1973). If, as seems likely, the X chromosomes of the females form chiasmata at meiosis, these species would constitute further examples of recombination being restricted in the heterogametic sex (Chapter 15).

While few animals reach the extremes shown by *W. venusta*, many have multiple sex chromosome systems which cause high proportions of the genome to display sex-linked inheritance. These multiple sex chromosomes arise through structural rearrangements involving sex chromosomes and autosomes. Structural rearrangements occasionally arise by spontaneous mutation, usually as a result of breakage and misrepair at interphase. Although they are likely to have no influence on sex determination and hence do not affect which sex is heterogametic, they do alter our perception of the numbers of chromosomes associated with sexual dimorphism. Chromosomes which are known to have recently altered in this way are sometimes referred to as neo-sex chromosomes. In practice, while we may suspect that a species has neo-sex chromosomes, we rarely have sufficient knowledge of its antecedents to be certain and so all sex-associated chromosomes have to be considered as sex chromosomes. In the Orthoptera (crickets, grasshoppers, locusts and mantids), for example, the majority of species have karyotypes consisting of telocentric chromosomes, heterogametic males with a single X chromosome (XO) and homogametic females with two X chromosomes (XX). The grasshopper *Oedaleonotus enigma* is unusual in having males with two distinct sex chromosomes, a metacentric represented in females and therefore known as an X chromosome and a telocentric restricted to males and known as a Y chromosome. The metacentric X chromosome is very likely to have been derived from Robertsonian fusion of a telocentric autosome with an ancestral telocentric X chromosome. The Y chromosome would then merely represent the homologue of the autosome fused to the old telocentric X chromosome (Figure 18.1). On these grounds, we suspect that in *Oe. enigma* we are dealing with a neo-X and a neo-Y. While X-A fusions can transform XX/ XO systems into XX/XY systems, they can also convert an XX/XY system into an XX/XY_1Y_2 system. By contrast, Y-A will convert an XX/XY system into an $X_1X_1X_2X_2$/X_1X_2Y system (Figure 18.1).

Some invertebrates appear to have undergone a series of Robertsonian fusions, resulting in the formation of large numbers of neo-sex chromosomes. The homogametic sex, being homozygous for X-A fusions, forms bivalents whereas the heterogametic sex, being heterozygous for X-A and Y-A fusions, forms interchange multivalents usually in alternate (zigzag) orientation. The interchange complex on the Y-A side of the multivalent constitutes a super-chromosomal differential segment, the only recombination with the X-A interchange complex taking place where chiasmata form at the ends of the chromosomes. Again, recombination is

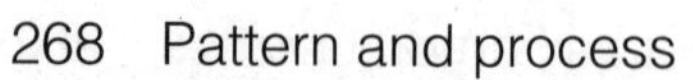

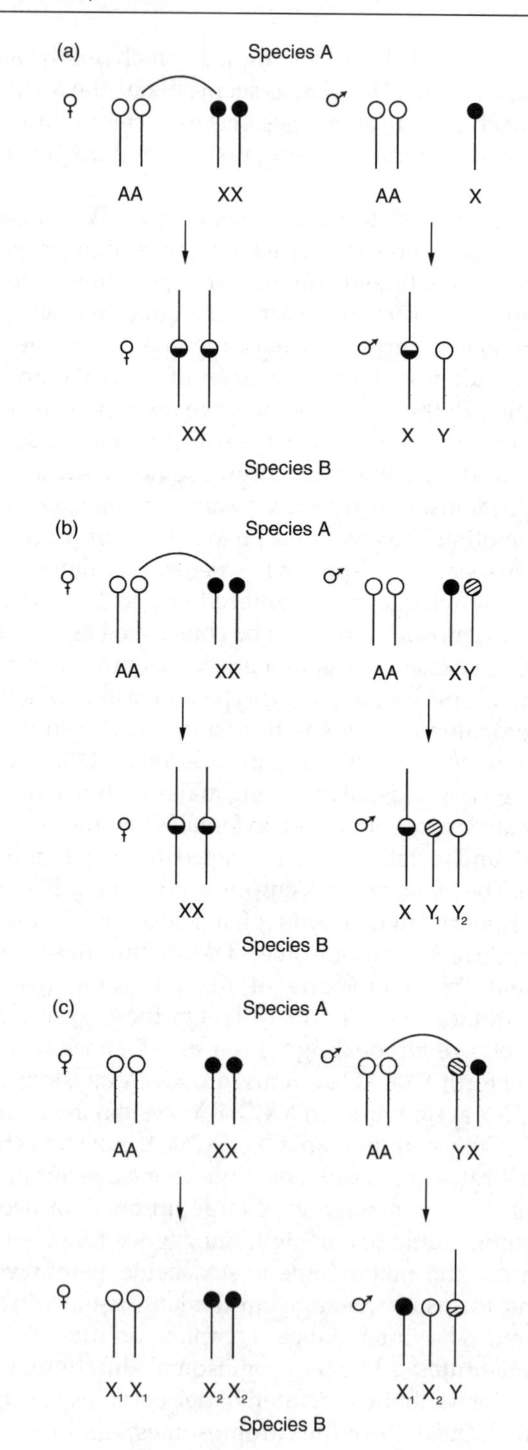
(a)
Species A
♀
AA
XX
♂
AA
X
♀
XX
♂
X Y
Species B
(b)
Species A
♀
AA
XX
♂
AA
XY
♀
XX
♂
X Y1 Y2
Species B
(c)
Species A
♀
AA
XX
♂
AA
YX
♀
X1 X1
X2 X2
♂
X1 X2 Y
Species B

restricted in the heterogametic sex. Interchange complexes involving sex chromosomes have been observed in heterogametic males of the centipede *Otocryptops sexspinosus* (♂$2n = 6 + X_{1\text{-}5}Y_{1\text{-}4}$), jumping spiders in the genus *Pellenes* (♂$X_{1\text{-}3}Y$) and in heterogametic females of the copepod *Diaptomus castor* (♀$X_{1\text{-}3}Y_{1\text{-}3}$). Cytological races of the Australian huntsman spider *Delena cancerides* exhibit distinctive patterns of Robertsonian fusion. One race has a full complement of telocentrics (♂$2n = 40 + X_{1\text{-}3}$) while another consists of 21 metacentrics and a single telocentric X_3 (♂$2n = 20 + X_{1+2}X_3$); in both races, the X chromosomes fail to pair but travel to the same pole at the first anaphase of meiosis. Three other races all include fusions between X chromosomes and autosomes and the males form multivalents of three ($X_{1\text{-}2}Y$), five ($X_{1\text{-}3}Y_{1\text{-}2}$) or nine chromosomes ($X_{1\text{-}5}Y_{1\text{-}4}$) at the first metaphase of meiosis. The true sex chromosomes in *D. cancerides* can be distinguished from the neo-sex chromosomes by their extensive blocks of constitutive heterochromatin (Rowell, 1985). Multiple sex chromosomes are common in many groups of arthropods, including crustaceans, spiders, assassin bugs (Reduviidae) and termites. Possibly the most extensive system of Robertsonian fusions involving sex chromosomes is found in Jamaican populations of the termite *Incisitermes schwarzi* where as many as 18 chromosomes are involved (♂$2n = 14 + X_{1\text{-}9}Y_{1\text{-}9}$).

18.5 SEX DETERMINATION

Sex linkage was discovered in 1906 by Doncaster and Raynor in the magpie moth *Abraxas grossulariata*. A pale mutant *lacticolor* was more frequent in females than in males yet females were never homozygous mutant. The phenomenon was explained by T.H. Morgan who found that white-eyed mutants of *Drosophila melanogaster* showed the reverse pattern, males being more often white-eyed than females. Morgan adopted E.B. Wilson's theory of chromosomal sex determination and proposed that white-eyed mutants were borne on the X chromosome which was present as a single copy in male *Drosophila* but as two copies in females. The *lacticolor* mutant in *Abraxas* was presumably also on the X chromosome but here females had one copy and males two. This contrasting pattern of sex determination has since been confirmed for many species in both orders of insects. The heterogametic sex is female in the Lepidoptera but male in the

Fig. 18.1 (opposite) Evolution of neo-sex chromosomes. (a) Fusion of an X chromosome with an autosome (A), in an XX/XO system, results in the formation of a new metacentric X with the unaffected autosome becoming a neo-Y. (b) X-A fusion in an XX/XY system results in a new metacentric X with the unaffected autosome behaving like a neo-Y (Y_2). The old Y remains (Y_1). (c) Fusion of an autosome with a Y chromosome results in the formation of a new metacentric Y, one arm of which pairs with the old X (X_1) and the other arm pairs with the unaffected autosome which behaves as a neo-X (X_2). Although the transition from one system to another is represented by a single arrow, it necessarily involves a series of crosses accompanied by selection in favour of the new metacentric chromosome.

Diptera. An analogous situation exists in vertebrates, the heterogametic sex being female in birds but male in mammals. The identity of the heterogametic sex appears to be strongly conserved, indicating that sex determination is more fundamental than sex chromosome evolution.

While analogous situations are apparent in invertebrates and vertebrates, the nature of sex determination is distinct. This distinction is highlighted by comparison of sex chromosome aneuploids in species with the same heterogametic sex. Aneuploids have irregular (although often stable) chromosome numbers, where some chromosome types are either under- or over-represented. Thus, XO individuals with a single X chromosome and no Y are fertile males in *Drosophila* but sterile females (Turner's syndrome) in humans. In the same way, XXY fruit flies are fertile females whereas XXY humans are infertile males (Klinefelter's syndrome). The Y chromosome has no bearing on the sex of *Drosophila* whereas it is the sole determinant of sex in humans. Sex in *Drosophila* is determined by the balance between the number of X chromosomes and the number of autosomes. One X or fewer per diploid set of autosomes results in a male whereas two Xs result in a female; intermediate numbers such as XX triploids produce intersexes. A similar system has been described for the nematode *Caenorhabditis elegans*, although the mode of gene action is distinct. The *Drosophila* system or some modification of it probably prevails in the majority of invertebrates, regardless of which sex is heterogametic. Examples of species where the male is heterogametic and XO are found in a wide range of invertebrates including nematodes, crustaceans, insects and spiders; heterogametic XO females are less common but have been reported in the Lepidoptera and in some species of copepod (White, 1973).

Clues to the nature of the sex-determining mechanism are also provided by contrasting patterns of chromosomal expression or suppression of gene activity. Thus, the single X chromosome in male *Drosophila* is more diffuse than either X in females. It is over-expressed to compensate for its single status. The opposite type of dosage compensation is found in orthopteran insects where the single X remains condensed at interphase, and its genes inactive. Evidently, X-linked genes are essential in male fruit flies but not in male grasshoppers. In mammals, one X is inactivated in females so that a single X is expressed in each sex. The pattern of inactivation is random in placental mammals but paternal in marsupials. Dosage compensation seems to be absent from many organisms where the female is heterogametic, including butterflies and birds. The nature of sex determination in birds is not so amenable to study as it is in mammals, largely due to the inviability of sex chromosome aneuploids. Studies of triploid chickens suggest that sex is determined by the presence of the W chromosome rather than the ratio of Z chromosomes to autosomes. Nevertheless, the number of Z chromosomes does have an effect, probably due to the lack of dosage compensation. Triploids with the constitution ZZZ are male but sterile, probably due to irregular chromosome segregation at meiosis, while those with ZZW begin as females but become sterile intersexes with the onset of sexual maturity.

The Y chromosome in humans carries a testis-determining factor, the TDF gene, which causes testes to develop and interrupt an otherwise female path of embryonic development. The TDF gene is the sole determinant of sex in humans: its presence causes XX individuals to be male and its absence causes an XY individual to be female. Searches for the TDF gene originally concentrated on the zinc-finger gene on the Y chromosome (ZFY). ZFY codes for a protein with multiple zinc-bearing sites which have strong DNA-binding activity and has the properties needed for activating transcription. It is found on the Y chromosomes of a wide range of placental mammals but is autosomal in marsupials. Moreover, XX human males can develop testes in the absence of ZFY. The next candidate gene was called SRY (sex-determining region of the Y chromosome) in case it too proved not be the TDF gene. In fact, the evidence in favour of SRY being equivalent to TDF is very strong. SRY is on the Y chromosome of placental mammals and marsupials and is essential for testis development. Surprisingly, SRY shows relatively low sequence homology, not only between placental mammals and marsupials but also within these groups (Foster *et al.*, 1992). Evidently, while the system of sex determination has been conserved, the gene responsible has not.

Sex-determining genes must exert their effects early in the development of an embryo. The gene which acts soonest will be the one which actually determines the sex of the embryo by setting in train a particular pattern of development. During the course of major evolutionary change, the old sex-determining genes may either act too late or their effects may no longer be of developmental importance. In such circumstances, novel sex-determining genes are likely to arise by mutation. Because it is not associated with the old sex-determining system, a new mutant will probably arise on an autosome. Like most new mutants, it will possibly involve some breakdown in function and hence be recessive. The affected pair of autosomes will now give rise to new sex chromosomes. The old sex chromosomes will revert to autosomal status; this is, of course, only possible in the homogametic sex. Under this scheme, therefore, novel sex-determining genes evolving in response to major evolutionary events arise in the homogametic sex which then becomes the new heterogametic sex. A possible mechanism for such change is suggested in Figure 18.2. It is based on known cases of sex-reversing mutants in the guppy *Lebistes reticulatus* (Winge, 1934).

The pattern of sex chromosome evolution and sex determination in the vertebrates is very striking. While some fish in the genera *Bathylagus* and *Fundulus* have males with large X and small Y chromosomes, the majority

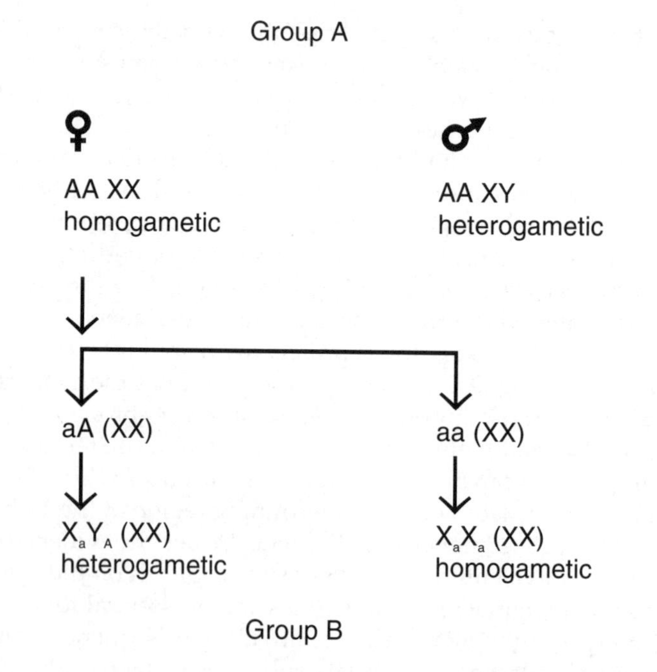

Fig. 18.2 Possible mechanism of reversing the heterogametic sex, following the arrival of a new sex-reversing mutant in an autosome. During the process, the mutant autosomes become the new sex chromosomes while the old homogametic sex chromosomes revert to autosomal status. Such mutants are commonly observed in fish notably the guppy *Lebistes reticulatus*. After Winge (1934).

of fish, such as *Lebistes*, have undifferentiated sex chromosomes and frequent sex-reversing mutants. One freshwater species, the swordtail *Xiphophorus maculatus*, is polymorphic for the heterogametic sex over much of its range in Mexico, both homo- and heterogametic males and females coexisting in the same populations. In contrast with the consistent patterns in birds and mammals, amphibians and reptiles provide examples of species with heterogametic males and others with heterogametic females. The W chromosome is female-determining in the axolotl *Ambystoma mexicanum*, in toads and in snakes whereas the Y chromosome is male-determining in salamanders, frogs and iguanas. In addition, some lizards, most turtles and all crocodilians show temperature-dependent sex determination. In *Alligator missisipiensis*, for example, eggs incubated at 33°C hatch into males whereas those incubated at 30°C hatch into females. The consistent systems in birds and in mammals are likely to have evolved from different groups of reptilian ancestors. Whether their early evolution involved the reversal of the heterogametic sex, along the lines of Winge's hypothesis (Figure 18.2), or whether it simply perpetuated that of the ancestors, remains to be determined. It is clear, however, that the sex-determining genes are likely to have evolved along separate lines, even if they prove to have similar molecular properties.

18.6 CONCLUSION

The great diversity of breeding systems exhibited by living organisms reveals clear-cut patterns. Each major taxonomic group is characterized by its own distinctive system which seems to have been strongly conserved and reinforced by the biological properties of the organisms concerned. In general, the effectiveness of the breeding system is correlated with the newness of the group and its morphological and physiological sophistication. Major biological innovations have been accompanied by significant improvements in breeding system, whose evolution proceeds over very long time periods due to slow but inexorable selective pressures. Species come into being and become extinct, but only occasionally does the speciation process affect the breeding pattern. This time scale has allowed complex interacting genetic systems to develop.

In a trail-blazing essay on amphimixis, August Weismann (1892) recognized the importance of sexual reproduction as the source of variation which Darwin had been looking for to counter-balance the loss of variation through natural selection. His thesis was strongly influenced by studies of chromosome behaviour during sexual reproduction in the horse nematode *Parascaris equorum* and the rejuvenation of cultures of *Paramecium* following conjugation. Weismann clearly recognized the potential of sexual reproduction for recombining genetic material and its implications for evolution.

18.7 SUMMARY

Our survey of the evolution of breeding systems in plants and animals has revealed ever more sophisticated systems for limiting inbreeding. Each evolutionary lineage is predetermined to some degree by the basic biological properties of the organisms concerned. Thus hermaphroditism tends to be self-perpetuating in most plants and some major groups of animals. The effectiveness of self-incompatibility systems in fungi is restricted by their mating behaviour which involves interactions between gametophytic tissues. Self-incompatibility systems are scarcely represented in lower plants but have become highly successful in flowering plants, due to the enclosure of the ovule and the evolution of the style and stigma. In contrast to the passive mating behaviour of plants, molluscs exhibit a wide range of active mating patterns. The majority of animals have bisexual species, with sex being determined by specific genes on specific chromosomes.

Genes associated with sexual dimorphism often become linked with those determining sex. Such linkage may take the form of multiple neo-sex chromosome systems or genetically isolated differential segments. The very segments which have evolved to promote recombination are themselves protected from it. The accumulation of mildly deleterious mutants in the differential segments is offset by methylation of the DNA and inactivation of the chromatin.

As with the arrival of the distinctive properties of flowering plants, so the vertebrates seem to have acquired a novel mode of sex-determining system. In contrast with the invertebrates where sex is determined by the balance between the number of sex chromosomes and the number of autosomes, sex in vertebrates is determined by the presence or absence of a sex-determining gene on a specific sex chromosome which is only found in one sex: the heterogametic sex. While the identity of the heterogametic sex is variable in the lower vertebrates, it is highly conserved in birds and mammals. Birds and mammals are likely to have unrelated sex-determining genes, although these may well have similar molecular properties.

18.8 FURTHER READING

Bell, G. (1988) *Sex and Death in Protozoa. The History of an Obsession.*
Darlington, C.D. (1978) *The Little Universe of Man.*
John, B. and Lewis, K.R. (1975) *Chromosome Hierarchy.*
John, B. and Miklos, G. (1987) *The Eukaryotic Genome in Development and Evolution.*
Lewis, D. (1979) *Sexual Incompatibility in Plants.*
Maynard Smith, J. (1978) *The Evolution of Sex.*
Read, K.C. and Marshall Graves, J.A. (1993) *Sex Chromosomes and Sex-determining Genes.*
Rees, H. and Jones, R.N. (1977) *Chromosome Genetics.*
Richards, A.J. (1997) *Plant Breeding Systems.*
Williams, G.C. (1975) *Sex and Evolution.*

References

Altman, P.H. and Dittner, D.S. (eds) (1972) *Biology Data Book*, Vol. I. Federation of American Societies for Experimental Biology, Bethesda, MD.

Arnold, M.L., Shaw, D.D. and Conteras, N. (1987) Ribosomal RNA-encoding DNA introgression across a narrow hybrid zone between two subspecies of grasshopper. *Proc. Natl. Acad. Sci., USA*, **84**, 3446–3450.

Bell, G. (1974) The reduction of morphological variation in natural populations of smooth newt larvae. *J. Anim. Ecol.*, **43**, 115–128.

Berry, R.J. and Crothers, J.H. (1968) Stabilizing selection in the dogwhelk (*Nucella lapillus*). *J. Zool. Lond.*, **155**, 5–17.

Bush, G.L. (1975) Modes of animal speciation. *Ann. Rev. Ecol. Syst.*, **6**, 334–364.

Cain, A.J. (1977) The uniqueness of the polymorphism of *Cepaea* (Pulmonata: Helicidae) in western Europe. *J. Conch.*, **29**, 129–136.

Cain, A.J. (1983) Ecology and ecogenetics of terrestrial molluscan populations. In: Russell-Hunter, W.D. (ed.), *The Mollusca, Vol. 6, Ecology*. Academic Press, New York, pp. 597–647.

Cain, A.J. (1988) The colours of marine bivalve shells with special reference to *Macoma balthica*. *Malacologia*, **28**, 289–318.

Callow, R.S. and Parker, J.S. (1985) Heirarchical control of multivalent formation in natural populations of a hexaploid perennial grass (*Koeleria vallesiana*). *Proc. R. Soc. Lond. B*, **223**, 459–473.

Chambers, G.K. (1988) The *Drosophila* alcohol dehydrogenase gene-enzyme system. *Adv. Genet.*, **25**, 39–107.

Charles, A.H. (1964) Differential survival of plant types in swards. *J. Brit. Grassland Soc.*, **19**, 198–204.

Clarke, B., Arthur, W., Horsley, D.T. and Parkin, D.T. (1978) Genetic variation and natural selection in pulmonate snails. In: Fretter, V. and Peake, J. (eds), *The Pulmonates, Vol. 2A, Systematics, Evolution and Ecology*. Academic Press, New York, pp. 220–270.

Clarke, C.A. and Sheppard, F.M. (1971) Further studies on the genetics of the mimetic butterfly *Papilio memnon* L. *Phil. Trans. R. Soc. B*, **263**, 35–70.

Cook, L.M. (1978) Zaire butterflies and faunal diversity in the tropics. *Biol. J. Linn. Soc.*, **10**, 347–360.

Cook, L.M. (1998) A two-stage model for *Cepaea* polymorphism. *Phil. Trans. R. Soc. Lond. B*, **353**, 1577–1593.

Daday, H. (1954) Gene frequency in wild populations of *Trifolium repens*. I. Distribution by latitude. *Heredity*, **8**, 61–78.

Darlington, C.D. (1973) *Chromosome Botany and the Origins of Cultivated Plants.* Allen and Unwin, London.
Dobzhansky, Th. (1971) Evolutionary oscillations in *Drosophila pseudoobscura*. In: Creed, E.R. (ed.) *Ecological Genetics and Evolution*. Blackwell, Oxford.
Ennos, R.A. (1982) Association of the cyanogenic loci in white clover. *Genet. Res.*, **40**, 65–72.
Ennos, R.A. (1983) Maintenance of genetic variation in plant populations. *Evol. Biol.*, **16**, 129–155.
Erwin, T.L. (1983) Beetles and other insects of tropical forest canopies at Manaus, Brazil, sampled by insecticidal fogging. In: Sutton, S.L., Whitmore, T.C. and Chadwick, A.C. (eds), *Tropical Rain Forest: Ecology and Management*. Blackwell, Oxford.
Ferris, C., Callow, R.S. and Gray, A.J. (1992) Mixed first and second division meiotic restitution in male meiosis of *Hierochloeumlaut odorata* (L.) Beauv. (Holy Grass). *Heredity*, **69**, 21–31.
Flatz, G. (1987) Genetics of lactose digestion in humans. *Adv. Hum. Genet.*, **16**, 1–77.
Foster, J.W. *et al.* (1992) Evolution of sex determination and the Y chromosome: SRY-related sequences in marsupials. *Nature*, **359**, 531–533.
Fowler, R.G., Degnen, G.E. and Cox, E.C. (1974) Mutational specificity of a conditional *Escherischia coli* mutator, mutD5. *Mol. Gen. Genet.*, **133**, 179–191.
Goodhart, C.B. (1987) Why are some snails visibly polymorphic, and others not? *Biol. J. Linn. Soc.*, **31**, 35–57.
Goss, R.J. (1983) *Deer Antlers. Regeneration, Function and Evolution*. Academic Press, New York.
Grant, P.R. (1986) *Ecology and Evolution of Darwin's Finches*. Princeton University Press, Princeton, NJ.
Hair, J.B. (1956) Sub-sexual reproduction in *Agropyron*. *Heredity*, **10**, 129–160.
Heller, J. (1993) Hermaphroditism in molluscs. *Biol. J. Linn. Soc.*, **48**, 19–42.
Hodgkin, J. (1990) Sex determination compared in *Drosophila* and *Caenorhabolitis*. *Nature*, **344**, 721–728.
Humphreys, M.O. and Nicholls, M.K. (1984) Relationships between tolerance to heavy metals in *Agrostis capillaris* L. (*A. tenuis* Sibth.). *New Phytol.*, **98**, 177–190.
Jones, J.S., Leith, B.H. and Rawlings, P. (1977) Polymorphism in *Cepaea*: a problem with too many solutions? *Ann. Rev. Ecol. Syst.*, **8**, 109–143.
Keast, A. (1969) Evolution of mammals on southern continents. VII. Comparisons of the contemporary mammalian faunas of the southern continents. *Quart. Rev. Biol.*, **44**, 121–167.
Kimura, M. (1955) Solution of a process of random genetic drift with a continuous model. *Proc. Natl. Acad. Sci.*, **41**, 144–150.
Lack, D. (1971) *Ecological Isolation in Birds*. Blackwell, Oxford.
Leary, R.F., Allendorf, F.W. and Knudson, K.L. (1983) Developmental stability and enzyme heterozygosity in rainbow trout. *Nature*, **301**, 71–72.
Legg, G. (1978) A note on the diversity of world Lepidoptera (Rhopalocera). *Biol. J. Linn. Soc.*, **10**, 343–347.
Leonard, J.L. (1990) The hermaphrodite's dilemma. *J. Theor. Biol.*, **147**, 361–372.
Lewontin, R.C. (1974) *The Genetic Basis of Evolutionary Change*. Colombia University Press, New York.
Linney, R., Barnes, B.W. and Kearsey, M.J. (1971) Variation for metrical characters in *Drosophila* populations. III. The nature of selection. *Heredity*, **27**, 163–174.

McNeilly, T. (1968) Evolution in closely adjacent plant populations. III. *Agrostis tenuis* on a small copper mine. *Heredity*, **23**, 99–108.
McNeilly, T. and Antonovics, J. (1968) Evolution in closely adjacent plant populations. IV. Barriers to gene flow. *Heredity*, **23**, 205–218.
Mallet, J. (1995) A species definition for the Modern Synthesis. *Trends Ecol. Evol.*, **10**, 294–299.
Mather, K. and Jinks, J.L. (1971) *Biometrical Genetics*. Chapman and Hall, London.
Meggers, B.J., Ayensu, E.S. and Duckworth, D.W. (eds) (1973) *Tropical Forest Ecosystems in Africa and South America: A Comparative Review*. Smithsonian Institution, Washington, DC.
Morgan, T.H. (1911) The application of the conception of pure lines to sex-limited inheritance and to sexual dimorphism. *Am. Nat.*, **45**, 65–78.
Moritz, C. (1984) Parthenogenesis in the endemic Australian lizard *Heteronotia binoei* (Gekkonidae). *Science*, **220**, 735–737.
Munroe, E. (1961) The classification of the Papilionidae (Lepidoptera). *Can. Entomol. (Suppl.)*, **17**, 1–51.
Nevo, E., Beiles, A. and Ben-Schlomo, R. (1984) The evolutionary significance of genetic diversity: ecological, demographic and life history correlates. In: Mani, G.S. (ed.), *Evolutionary Dynamics of Genetic Diversity*. Springer, Berlin.
Ochman, H., Jones, J.S. and Selander, R.K. (1987) Large scale patterns of genetic differentiation at enzyme loci in the land snails *Cepaea nemoralisi* and *Cepaea hortensis*. *Heredity*, **58**, 127–138.
Odum, E.P. (1959) *Fundamentals of Ecology*, Saunders, Philadelphia, PA.
Parker, J.S., Jones, G.H., Tease, C. and Palmer, R.W. (1976) Chromosome-specific control of chiasma formation in *Hypochoeris* and *Crepis*. In: Jones, K. and Branham, P.E. (eds), *Current Chromosome Research*. North-Holland, Amsterdam.
Patterson, C. (1980) Cladistics. *Biologist*, **27**, 234–240.
Pooni, H.S. and Jinks, J.L. (1981) The true nature of the non-allelic interactions in *Nicotiana rustica* revealed by association crosses. *Heredity*, **47**, 389–400.
Rees, H. and Dale, P.J. (1974) Chiasmata and variability in *Lolium* and *Festuca* populations. *Chromosoma*, **47**, 335–351.
Reid, D.G. (1986) *The Littorinid Molluscs of Mangrove Forests in the Indo-Pacific Region: the Genus Littoraria*. British Museum (Natural History), London.
Reid, D.G. (1987) Natural selection for apostasy and crypsis acting on the shell colour polymorphism of a mangrove snail, *Littoraria Filosa* (Sowerby) (Gastropoda: Littorinidae). *Biol. J. Linn. Soc.*, **30**, 1–24.
Rowell, D.M. (1985) Complex sex-linked fusion heterozygosity in the Australian huntsman spider *Delena cancerides* (Araneae: sparassidae) *Chromosoma*, **93**, 169–176.
Selander, R.K. and Ochman, H. (1983) The genetic structure of populations as illustrated by molluscs. *Curr. Topics Biol. Med. Res.*, **10**, 93–123.
Shaw, D.D. (1974) Genetic and environmental components of chiasma control. III. Genetic analysis of chiasma frequency variation in two selected lines of *Schistocerca gregaria*. *Chromosoma*, **46**, 365–375.
Shorrocks, B. and Rosewell, J. (1987) Spatial patchiness and community structure: coexistence and guild size of drosophilids on ephemeral resources. In: Gee, J.H.R. and Giller, P.S. (eds), *Organization of Communities Past and Present*. Blackwell, Oxford.
Sokal, R.R. and Sneath, P.H.A. (1963) *Principles of Numerical Taxonomy*. Freeman, San Francisco, CA.

Symeonides, L., McNeilly, T. and Bradshaw, A.D. (1985) Differential tolerance of three cultivars of *Agrostis capillaris* to cadmium, copper, lead, nickel and zinc. *New Phytol.*, **101**, 309–315.

Thoday, J.M. (1979) Polygene mapping: uses and limitations. In: Thompson, J.M. and Thoday, J.M. (eds), *Quantitative Genetic Variation*. Academic Press, New York.

Van Valen, L. and Mellin, G.W. (1967) Selection in natural populations. 7. New York babies (fetal life study). *Ann. Hum. Genet.*, **31**, 109–127.

Weismann, A. (1892) Amphimixis or the essential meaning of conjunction and sexual reproduction. In: Poulton, E.B. and Shipley, A.E. (eds), *Essays upon Heredity and Kindred Biological Problems*, Vol. II. Oxford University Press, Oxford.

White, M.J.D. (1973) *Animal Cytology and Evolution*. Cambridge University Press, Cambridge.

Whittaker, R.H. and Margulis, L. (1978) Protist classification and the kingdoms of organisms. *Biosystems*, **10**, 3–18.

Williams, D.M. and Embley, T.M. (1996) Microbial diversity: domains and kingdoms. *Ann. Rev. Ecol. Syst.*, **27**, 569–595.

Williamson, P.G. (1981) Palaeontological documentation of speciation in Cenozoic molluscs from Turkana Basin. *Nature*, **293**, 437–443.

Winge, Ö (1934) The experimental alteration of sex chromosomes into autosomes and vice versa, as illustrated by *Lebistes*. *C.R. Trav. Lab. Carlsberg*, **21**, 1–49.

Wright, S. and Dobzhansky, Th. (1946) Genetics of natural populations. XII. Experimental reproduction of some of the changes caused by natural selection in certain populations of *Drosophila pseudoobscura*. *Genetics*, **31**, 125–256.

Yosida, T.H., Tsuchiya, K. and Moriwaki, K. (1971) Karyotypic difference of black rats, *Ratus ratus*, collected in various localities of east and southeast Asia and Oceania. *Chromosoma*, **33**, 252–267.

Zarchi, Y., Simchen, G., Hillel, J. and Schaap, T. (1972) Chiasmata and the breeding system in wild populations of diploid oats. *Chromosoma*, **38**, 77–94.

Zouras, E. and Foltz, D.W. (1987) Isozymes. *Curr. Topics Biol. Med. Res.*, **13**, 1–59.

Bibliography

Arthur, W. (1987) *The Niche in Competition and Evolution*. Wiley, Chichester.
Avise, W. (1987) *Molecular Markers, Natural History and Evolution*. Chapman and Hall, London.
Begon, M., Mortimer, M. and Thompson, D.J. (1996) *Population Ecology*. Blackwell, Oxford.
Bell, G. (1988) *Sex and Death in Protozoa. The History of an Obsession*. Cambridge University Press, Cambridge.
Berg, P. and Singer, M. (1992) *Dealing with Genes. The language of Heredity*. University Science Books, Mill Valley, CA.
Berry, R.J. (1977) *Inheritance and Natural History*. Collins, London.
Berryman, A. (1999) *Principles of Population Dynamics and their Application*. Stanley Thornes, Cheltenham.
Bradshaw, A.D. and McNeilly, T. (1981) *Evolution and Pollution*. Arnold, London.
Cavalli-Sforza, L.L. and Bodmer, W.F. (1971) *The Genetics of Human Populations*. Freeman, San Francisco, CA.
Cavalli-Sforza, L.L., Menozzi, P. and Piazza, A. (1994) *The History and Geography of Human Genes*. Princeton University Press, Princeton, NJ.
Charlesworth, B. (1980) *Evolution in Age-structured Populations*. Cambridge University Press, Cambridge.
Claridge, M.F., Dowah, H.A. and Wilson, M.R. (eds) (1997) *Species. The Units of Diversity*. Chapman and Hall, London.
Crow, J.F. (1986) *Basic Concepts in Population, Quantitative and Evolutionary Genetics*. Freeman, New York.
Crow, J.F. and Kimura, M. (1970) *An Introduction to Population Genetics Theory*. Harper and Row, New York.
Darlington, C.D. (1978) *The Little Universe of Man*. Allen and Unwin, London.
Diamond, J. and Case, T.J. (eds) (1986) *Community Ecology*. Harper and Row, New York.
Eldredge, N. (1985) *Unfinished Synthesis*. Oxford University Press, Oxford.
Endler, J.A. (1986) *Natural Selection in the Wild*. Princeton University Press, Princeton, NJ.
Falconer, D.S. and Mackay, T.F.C. (1996) *Introduction to Quantitative Genetics*. Longman, Harlow.
Ford, E.B. (1971) *Ecological Genetics*. Chapman and Hall, London.
Futuyma, D.J. (1986) *Evolutionary Biology*. Sinauer, Sunderland, MA.
Gale, J.S. (1990) *Theoretical Population Genetics*. Unwin Hyman, London.

Gaston, K. and Spicer, J.I. (1998) *Biodiversity. An Introduction*. Blackwell, Oxford.
Georghiou, G.P. and Saito, T. (eds) (1983) *Pest Resistance to Pesticides*. Plenum, New York.
Gillespie, J.H. (1991) *The Causes of Molecular Evolution*. Oxford University Press, Oxford.
Grant, P.R. (1986) *Ecology and Evolution of Darwin's Finches*. Princeton University Press, Princeton, NJ.
Grant, P.R. (ed.) (1997) *Evolution on Islands*. Oxford University Press, Oxford.
Hanski, I.A. and Gilpin, M.E. (eds) (1997) *Metapopulation Biology. Ecology, Genetics and Evolution*. Academic Press, San Diego, CA.
Hartl, D.L. and Clark, A.G. (1989) *Principles of Population Genetics*. Sinauer, Sunderland, MA.
Hartl, D.L. and Jones, E.W. (1998) *Genetics: Principles and Analysis*. Jones and Bartlett, Sudbury, MA.
Hassell, M.P. (1978) *The Dynamics of Arthropod Predator – Prey Systems*. Princeton University Press, Princeton, NJ.
Hedrick, P.W. (1984) *Population Biology*. Jones and Bartlett, Boston, MA.
John, B. (1976) *Population Cytogenetics*. Arnold, London.
John, B. and Lewis, K.R. (1975) *Chromosome Hierarchy*. Oxford University Press, Oxford.
John, B. and Miklos, G. (1987) *The Eukaryotic Genome in Development and Evolution*. Allen and Unwin, London.
Kearsey, M.J. and Pooni, H.S. (1996) *The Genetical Analysis of Quantitative Traits*. Chapman and Hall, London.
Kimura, M. (1983) *The Neutral Theory of Molecular Evolution*. Cambridge University Press, Cambridge.
King, M. (1993) *Species Evolution. The Role of Chromosome Change*. Cambridge University Press, Cambridge.
Krebs, C.J. (1994) *Ecology*. Harper and Row, New York.
Korol, A.B., Preygel, I.A. and Preygel, S.I. (1994) *Recombination Variability and Evolution*. Chapman and Hall, London.
Lewin, B. (1997) *Genes VI*. Oxford University Press, Oxford.
Lewis, D. (1979) *Sexual Incompatibility in Plants*. Arnold, London.
Lewontin, R.C. (1974) *The Genetic Basis of Evolutionary Change*. Columbia University Press, New York.
Li, W.-H. (1997) *Molecular Evolution*. Sinauer, Sunderland, MA.
Li, W.-H. and Graur, D. (1991) *Fundamentals of Molecular Evolution*. Sinauer, Sunderland, MA.
Majrus, M.E.N. (1998) *Melanism. Evolution in Action*. Oxford University Press, Oxford.
Manly, B. F. J. (1985) *The Statistics of Natural Selection in Animal Populations*. Chapman and Hall, London.
Mather, K. (1973) *Genetical Structure of Populations*. Chapman and Hall, London.
Mather, K. and Jinks, J.L. (1971) *Biometrical Genetics*. Chapman and Hall, London.
May, R. M. (1973) *Stability and Complexity in Model Ecosystems*. Princeton University Press, Princeton, NJ.
May, R.M. (ed.) (1981) *Theoretical Ecology*. Blackwell, Oxford.
Maynard Smith, J. (1978) *The Evolution of Sex*. Cambridge University Press, Cambridge.
Maynard Smith, J. (1998) *Evolutionary Genetics*. Oxford University Press, Oxford.

Mayr, E. (1970) *Populations, Species and Evolution*. Harvard University Press, Cambridge, MA.
Mayr, E. (1976) *Evolution and the Diversity of Life*. Harvard University Press, Cambridge, MA.
Mayr, E. and Provine, W.B. (eds) (1980) *The Evolutionary Synthesis*. Harvard University Press, Cambridge, MA.
Mitton, J.B. (1998) *Selection in Natural Populations*. Oxford University Press, Oxford.
National Research Council (1986) *Pesticide Resistance. Strategies and Tactics for Management*. National Academy Press, Washington, DC.
Nei, M. (1987) *Molecular Evolutionary Genetics*. Columbia University Press, New York.
Nei, M. and Koehn, R.K. (eds) (1983) *Evolution of Genes and Proteins*. Sinauer, Sunderland, MA.
Ohta, T. (1980) *Evolution and Variation of Multigene Families*. Springer, Berlin.
Ohta, T. and Aoki, K. (eds) (1985) *Population Genetics and Molecular Evolution*. Japan Scientific Societies Press, Tokyo.
Read, K.C. and Marshall Graves, J.A. (1993) *Sex Chromosomes and Sex-determining Genes*. Harwood, Chur, Switzerland.
Rees, H. and Jones, R.N. (1997) *Chromosome Genetics*. Arnold, London.
Richards, A.J. (1997) *Plant Breeding Systems*. Thornes, London.
Ricklefs, R.E. (1997) *The Economy of Nature: a Textbook of Basic Ecology*. Freeman, San Francisco, CA.
Ridley, M. (1993) *Evolution*. Blackwell, Oxford.
Roff, D.A. (1997) *Evolutionary Quantitative Genetics*. Chapman and Hall, London.
Roughgarden, I. (1979) *Theory of Population Genetics and Evolutionary Ecology: An Introduction*. Macmillan, New York.
Royama, T. (1992) *Analytical Population Dynamics*. Chapman and Hall, London.
Sheppard, P.M. (1975) *Natural Selection and Heredity*. Hutchinson, London.
Skelton, P. (ed.) (1996) *Evolution. A Biological and Palaeontological Approach*. Addison-Wesley, Harlow.
Wallace, B. (1970) *Genetic Load*. Prentice-Hall, Englewood Cliffs, NJ.
Wallace, B. (1991) *Fifty Years of Genetic Load. An Odyssey*. Cornell University Press, Ithaca, NY.
Watson, J.D, Hopkins, N.H, Roberts, J.W. *et al.* (1987) *The Molecular Biology of the Gene*. Benjamin/Cummings, Menlo Park, CA.
Weaver, R.F. and Hedrick, P.W. (1997) *Genetics*. W.C. Brown, Dubuque, IA.
White, M.J.D. (1978) *Modes of Speciation*, Freeman, San Francisco, CA.
Williams, G.C. (1975) *Sex and Evolution*. Princeton University Press, Princeton, NJ.
Williamson, M. (1972) *The Analysis of Biological Populations*. Arnold, London.
Williamson, M. (1981) *Island Populations*. Oxford University Press, Oxford.
Wilson, E.O. (ed.) (1988) *Biodiversity*. National Academy Press, Washington, DC.
Wright, S. (1978) *Evolution and the Genetics of Populations, Vol. 4, Variability Within and Among Natural Populations*. University of Chicago Press, Chicago.

Index